SCIENCE, PHILOSOPHY
AND MUSIC

DE DIVERSIS ARTIBUS

COLLECTION DE TRAVAUX
DE L'ACADÉMIE INTERNATIONALE
D'HISTOIRE DES SCIENCES

COLLECTION OF STUDIES
FROM THE INTERNATIONAL ACADEMY
OF THE HISTORY OF SCIENCE

DIRECTION
EDITORS

EMMANUEL
POULLE

ROBERT
HALLEUX

TOME 63 (N.S. 26)

BREPOLS

PROCEEDINGS OF THE XXth INTERNATIONAL CONGRESS
OF HISTORY OF SCIENCE (Liège, 20-26 July 1997)

VOLUME XX

SCIENCE, PHILOSOPHY AND MUSIC

Edited by

Erwin NEUENSCHWANDER and Laurence BOUQUIAUX

BREPOLS

The XXth International Congress of History of Science was organized by the Belgian National Committee for Logic, History and Philosophy of Science with the support of :

ICSU
Ministère de la Politique scientifique
Académie Royale de Belgique
Koninklijke Academie van België
FNRS
FWO
Communauté française de Belgique
Région Wallonne
Service des Affaires culturelles de la Ville de Liège
Service de l'Enseignement de la Ville de Liège
Université de Liège
Comité Sluse asbl
Fédération du Tourisme de la Province de Liège
Collège Saint-Louis
Institut d'Enseignement supérieur "Les Rivageois"

Academic Press
Agora-Béranger
APRIL
Banque Nationale de Belgique
Carlson Wagonlit Travel - Incentive Travel House

Chambre de Commerce et d'Industrie de la Ville de Liège
Club liégeois des Exportateurs
Cockerill Sambre Group
Crédit Communal
Derouaux Ordina sprl
Disteel Cold s.a.
Etilux s.a.
Fabrimétal Liège - Luxembourg
Generale Bank n.v. - Générale de Banque s.a.
Interbrew
L'Espérance Commerciale
Maison de la Métallurgie et de l'Industrie de Liège
Office des Produits wallons
Peeters
Peket dè Houyeu
Petrofina
Rescolié
Sabena
SNCB
Société chimique Prayon Rupel
SPE Zone Sud
TEC Liège - Verviers
Vulcain Industries

D/2002/0095/59
ISBN 2-503-51414-6
Printed in the E.U. on acid-free paper

TABLE OF CONTENTS

Part one
SCIENTIFIC MODELS FROM ANTIQUITY TO THE PRESENT

Part two
SCIENCE AND PHILOSOPHY

Part three
SCIENCE AND MUSIC

SCIENTIFIC MODELS
FROM ANTIQUITY TO THE PRESENT

THE MATHEMATICAL MODEL : EPISTEMOLOGICAL TOOL OR IDEOLOGICAL NOTION ?

Martin ZERNER

We sum up here the methodology and some of the conclusions of a research project which has been proceeding for a number of years.

The methodology is twofold. One part is to make a historical study focusing on the appearance of the *word* " model " in scientific contexts. This makes it possible *not* to rely on a preconceived idea of what a mathematical model is and to detect a great variety of meanings. The other part is to make case studies of models in different fields in order to compare the ways in which they work.

It is clear that the historical study cannot be exhaustive. On one hand every field and subfield cannot be explored. On the other hand, some occurrences certainly will escape the investigation. Nevertheless, if in a definite field no occurrence has been found, we can be reasonably sure that the word has been at most exceptionally used.

With this qualification, we first find the word " model " in the context of physics starting with Maxwell[1]. It was taken up by several authors, including Boltzmann and Hertz[2]. Maxwell's first model was geometrical. However, most of these models of physical phenomena were themselves physical systems the function of which was to give the theory of the phenomenon of which they were a model. Rutherford's model of the atom (1911) is quite typical. Notice that there is no Bohr model, but, in his own words, a contribution to the theory of Rutherford's atomic model[3]. For this reason, I consider these models as *physical,* not *mathematical.* Whether they provided an explanation of the phys-

---

1. J.C. Maxwell, " On Faradays' Lines of Force ", *Phil. Trans. Camb. Soc.,* 10 (1857), part I.

2. See R. Müller, " Zur Geschichte des Modelldenkens und des Modellbegriffs ", in H. Stachowiak (ed.), *Modelle-Konstruktion der Wirklichkeit,* München, 1973. The author is grateful to Prof. Neuenschwander for this reference, the only one to his knowledge to pay attention to the actuel use of the word. Unfortunately, it does not cover the 20th century.

3. H. Bohr, " On the Constitution of Atoms and Molecules ", *Philosophical Magazine,* VI, 26, (1913), 1-25.

ical phenomenon or a simple fiction which helped the mind to find out its mathematics was controversial[4]. It is in the context of this controversy that Mach uses the phrase " mathematical model " for the atomic theory to emphasize his thesis according to which the atoms had no objective existence. Incidentally, this phrase was used in the same period with a quite different meaning, namely a concrete representation of a mathematical object, typically a plaster molding of a surface.

The situation was quite different in the 1920s and 1930s. The word " model " is then very seldom found in physics. This may be due to the fact that the experimental evidence for the existence of atoms and subatomic particles was now overwhelming. Also with the advent of quantum physics a particle can no longer be conceived with the intuitions which we worked out in the macroscopic world and its very description relies now on fairly advanced mathematics. As a consequence the distinction between a physical and a mathematical model is blurred. This is e.g., the status of what has been first called the Thomas-Fermi theory or approximation and later increasingly often the Thomas-Fermi model.

The year 1936 sees the permanent introduction of the word " model " in two fields : mathematical statistics[5] and economics, more precisely econometrics[6]. The first deserves a closer study. In economics the context of " model " is two-fold. At the beginnings of articles or books, we find general statements about mathematical models. But when the word appears in specific situations, the context is statistical. This remains true until 1944, the year of the appearence of Haavelmo's famous *manifesto*[7] which theorizes very clearly the necessity of a probabilistic mathematical formulation if an economic theory has to be checked against available data. It is impossible to sum up this article in a few lines, but let us try and give the main idea. The economist, contrary to e.g. the agricultural engineer, has no means whatsoever to test his theories against experimental situations in which he may choose various values of some of the parameters. He must rely on tests of hypotheses. Now these are meaningful only if there is a probabilistic theory to test (the phrases used by Havelmoo are " probabilistic model " and " theoretical model ").

A completely different point of view is expressed in that same year 1944 in von Neumann and Morgenstern's still more famous *Theory of Games and Eco-*

4. E. Bellone, *I Modelli e la Concezione del Mondo nella Fisica moderna da Laplace a Bohr*, Milano, 1973 gives a detailed account of the main developments in 19th century physics, focusing on this controversy.

5. J. Neyman, E. Pearson, " Contributions to the Theory of Testing Statistical Hypotheses ", *Statistical Research Memoirs*, vol. 1, London, 1936, 1-37. Reprinted in *Joint Statistical Papers*, Berkeley, 1967.

6. J. Tinbergen, " An Economic Policy for 1936 ", in J. Tinbergen (ed.), *Selected Papers*, Amsterdam, 1959.

7. T. Haavelmo, " The probability approach in econometrics ", *Econometrica*, 12 (1944), (supplement).

nomic Behavior[8]. There appears the first statement of the conception of the mathematical model as the general method of the sciences which is now so widespread. In view of this, a special study was devoted to von Neumann[9].

Von Neumann very clearly expressed his conception of models, indeed a conception of the sciences in general, in a later article about physics : " To begin, we must emphasize a statement which I am sure you have heard before but which must be repeated again and again. It is that the sciences do not try to explain, they hardly even try to interpret, they mainly make models. By a model is meant a mathematical construct which, with the addition of some verbal interpretations, describes observed phenomena "[10].

It must be added that in this conception " model " and " theory " are synonymous. The statement must be checked against the way in which von Neumann himself spoke about models in his specific scientific works. He uses the word in the context of three fields : logic, physics and economics. We need not insist on logic where it appears in one article of 1925 with its purely technical meaning[11].

The works of von Neumann in physics are clearly separated into two periods : 1927-1934 (quantum physics) and 1941-1959 (mostly fluid mechanics). In the first period, the word " model " is to be found thrice, in rather obscure sentences. In the second period, it appears in 5 papers out of 26[12]. It has in these papers a clear meaning. Some concrete situation is to be studied. Physics provides an unquestioned theory of this situation, only it is too complicated to be carried out to make concrete predictions. The model is a simpler situation in which this can be done and there are good reasons to believe that this simpler situation is an approximation of the genuine one. An exception is Burgers' model with which we cannot deal here. Contrary to " model ", " theory " is often used in both periods and one word is *opposed* to the other on several occasions. So the physicist von Neumann is in flat contradiction with von Neumann the essayist on physics.

Von Neumann published two works on economics. The first has nothing to do with the notion of model except that the word appears in the misleading title

8. J. von Neumann, O. Morgenstern, *Theory of Games and Economic Behavior*, Princeton, 1944.

9. M. Zerner, " Analogie et modèles chez von Neumann ", submitted to *Rivista per la Storia delle Scienze*.

10. J. von Neumann, " Method in the Physical Sciences ", in L. Leary (ed.), *The Unity of Knowledge*, New York, 1955. All the works of von Neumann mentioned here except his books are reprinted in J. von Neumann, *Collected Works*, Oxford, Londres, New York, Paris, 1961-1963 (6 vols).

11. J. von Neumann, " Eine Axiomatisierung der Mengenlehre ", *Journal für die reine und angewandte Mathematik*, 154 (1925), 219-240.

12. One should may be add a paper with the title " The Point Source Model ". Only the word does not appear elsewhere in the paper. This paper was only published after von Neumann's death in his *Collected Works* and the title may have been chosen by Taub who edited it. The name " point source model " had been adopted in stellar physics when the *Collected Works* appeared.

of the English translation (1945)[13]. The other is the joint book with Morgenstern. As has already been mentioned, von Neumann's conception of models and their function in the sciences is already expressed there, indeed repeatedly. It is put to work in two ways. In a general way, it enables the authors to justify the kind of mathematization they perform in economics by analogy with physics. The other way is quite specific, it justifies the use of a numerical utility (cardinal in the usual economic terminology) by a detailed comparison with the case of temperature. Both rely on the conception of the sciences of von Neumann the essayist.

Within a few years after the publication of this book, the use of the word " model " became widespread in economics, usually with the same meaning as in the book. The next step was its being taken up in the so called behavioral sciences. This can be dated to 1951-1953 and it is quite clear that it was an importation from economics[14].

Operational research appears as the next field to have taken up the word. It was not used in the military (and secret) childhood of the field as can be seen in the book of Morse and Kimball[15]. The first volumes of the journal *Operations Research,* which started in 1952, contain occasional occurrences. It becomes of current use in the second half of the 1950s.

Population dynamics is supposed to be a great user of models, but the word is to be found only in the late fifties and at that time in a statistical context only.

In applied mathematics, the word " model " appears around 1965 and becomes usual in the 1970s in the United States and a few years later in France. This is conspicuously late in view of its later overwhelming presence in the literature of the field. It must be added that its meanings are manifold and it happens that the same author uses two different ones within a few lines (probably without realizing it). It is of interest to observe that in general papers about the methods of applied mathematics some authors illustrate their view of what a model is by examples taken from their practice with technical applications based on physics adding *in fine* " this is also the way it works in economics ".

In physics, the evolution is very complicated. From, very roughly, the middle of the century on, physicists speak about " models " again in various sub-

13. " A Model of General Equilibrium ", the original reference is : J. von Neumann, " Ueber ein œkonomisches Gleichungssystem und eine Verallgemeinerung des Brouwerschen Fixtpunktsatzes ", in K. Menger (ed.), *Ergebnisse eines mathematischen Kolloquiums*, Vienna, 1938. Notice that the object of the paper is not what economists call general equilibrium.

14. M. Zerner, " Paul Lazarsfeld et la Notion de Modèle mathématique ", in J. Lautman, B.P. Lecuyer (eds), *Paul Lazarsfeld (1901-1976)*, Paris, 1998.

15. P. Morse, G. Kimball, *Methods of Operational Research*, Cambridge, New York, London, 1951.

fields and with different meanings. But the field is the very last to adopt the word " model " in the sense of von Neumann the essayist.

Now a few words about the three case studies which have been carried out ; others would of course be highly desirable. They had obviously to be chosen in quite different fields. One of these fields had to be physics or some technology based on physics. The latter was deemed preferable as more typical and also, why not admit it, more familiar to the author. The oil reservoir simulations, as they appeared in the engineers' literature of the early 1970s were chosen for this case study. Another obvious choice was economics, mainstream economics of course. Now, very different things are called " models " in economics. To mention three, we have the supposedly basic theoretical models among which Arrow-Debreu is the best known, the big econometric models for prediction and a large number of intermediate models which rely on the former to study specific questions. These last make up the bulk of the research literature in economics ; one of them was selected for this and other reasons, namely a model of borrowing for a developing country[16]. To compare with a field quite different from both the preceding ones, the famous Lotka-Volterra systems (later called models) for population dynamics were studied.

The study of each model has to be thorough enough to avoid superficial resemblances and it must make clear the general principles on which the models work in order to be able to make comparisons. This makes it impossible to report them here, even in summary. The conclusions they lead to confirm what appeared in the historical study, namely that models in physics and in economics rely on very different principles. The case of the Lotka-Volterra equations is different from both, moreover there is a big difference between the overly famous equations for a predator-prey system and the less well known but much more firmly grounded equations for two species in competition for food.

It can be safely concluded that the use of any variant of the von Neumann conception of models can only lead to epistemological confusion. The ideological consequences of this confusion still need investigation. However, the contrast between a technological model based on physics and a mainstream economic model is clear. The first one relies on the principles of physics and controlable approximations to give indications about what can be gained in terms of efficiency (and ultimately profit). The second one is supposed, in the dominant view, to be justified by its logical coherence and a comparison with reality. In fact, when it comes to specific questions, the logical coherence is obtained at the cost of fantastic extra assumptions and the comparison with reality is, to put it mildly, very weak. In the specific case studied, there is none. And the historical context of the model shows that its function was to justify

16. R. Dornbusch, " Real Interest Rates, Home Goods and Optimal External Borrowing ", *Journal of Political Economy*, 91 (1983), 141-153.

the policies of macroeconomic adjustment to which the IMF was switching at that time.

The Discernment of Ancient Modeling Efforts

Don Faust

In the history of science we see marches, many yet unfinished, toward a precise and complete description of Reality. To illustrate the great length of some of these marches, we consider three problem areas in the history of modeling in science.

First, considering man's early struggles to model the human reproductive process leads to revisiting, from another perspective, an *old conjecture* concerning the 260 count of ancient American civilizations. Second, considering man's early attempts to model the moon's periodicity leads to a *new conjecture* concerning the prominence of sixty in the ancient numeration systems of the Near East. And third, considering man's attempts to model the concept of negation leads to some precise mathematical logic machinery and a *theorem* clarifying a connection between part of Aristotle's work 2000 years ago on privatives and part of recent work by Artificial Intelligence (AI) researchers on negation. This note is by necessity brief : for more detail, the reader may request the first reference in regard to problems one and two, the second for problem three.

The Count of 260

The Tzolkin count of 260 of ancient American civilizations consists of consecutively forming pairings of two primitive numeral sets, one of 20 and another of 13. The first of one set is paired with the first of the other set, then the second of one with the second of the other, and so on, cycling through the sets repeatedly. The count is finished when all possible pairings have been formed. In this way one has a constructed numeral system, an ordered numeral set consisting of 260 unique pairs.

This invention, for generating a counter set of larger cardinality from two given smaller counter sets, is certainly ingenious. It antedates the later place-value system of base 20 in ancient America, and so can be seen to be a partial solution to the problem of construction of a system of numeration allowing, as place-value systems do, the counting of sets of arbitrarily large cardinality.

That such a counting system developed at all is, of course, interesting enough. But what is even more interesting is that at least one other ancient culture developed a system of similar design. The ancient Chinese developed a system based on such pairing as well, the Chia-tsu numeration system based on one set of 12 counters and another set of 10 counters, a sixty count of pairs.

Let us look at the design involved in these two systems. For example, for the two counter sets with numerals a,b,c,d and 1,2,3,4,5,6 respectively, the counter set generated by pairing from them would consist of the following sequence of twelve numerals (each one a pair formed by the juxtaposition of one numeral from each of the two given counter sets) :

$$a1, b2, c3, d4, a5, b6, c1, d2, a3, b4, c5, d6.$$

Note the uniqueness of each element (pair) of this derived counter set, giving a perfectly valid set of primitive numerals. In general, we may define such numeral sets as follows. *The Pairing-Generated Counter Set* generated by a counter set A with first element a and last element b and a counter set C with first element c and last element d is the counter set S such that :

1) the first element of S is ac ; and

2) if xy is in S then : xy is the last element of S in case x=b and y=d ; otherwise xy has a successor in S, formed with the successor of x and the successor of y if x<b and y<d, with a and the successor of y if x=b and y<d, and with the successor of x and c if x<b and y=d.

As is easily seen, the cardinality of any pairing-generated counter set is the least common multiple of the cardinalities of the two sets upon which it is based. The cardinality of the Tzolkin, based on sets of cardinalities 13 and 20, is indeed 260, while the cardinality of the Chia-tsu, based on sets of cardinalities 12 and 10, is sixty.

It would seem that the Chia-tsu's sixty is simply derivative, by pairing-generated counting, from the natural sets of 12 (an undercount of the number of lunations in a solar year) and the obvious 10. The Tzolkin's 260, however, remains enigmatic : the 20 is as obvious as the 10 of the Chia-tzu, but why the 13 set, and the consequent pairing-generated 260 ?

Although there have been numerous conjectures concerning the origin of the 260 count, it is unfortunate that the conjecture that it was originally developed in connection with the human gestation period, a conjecture made as early as 1895 by E. Forstemann[1], has been uniformly dismissed as untenable. Namely, it has been uniformly argued, as recently as 1960[2], that 260 is not

1. E. Seler, *et al.*, *Mexican and Central American Antiquities, Calendar Systems, and History*, Washington, DC, 1904, 532.

2. J.E. Thompson, *Maya Hieroglyphic Writing*, 1960, 98.

even close to the 280 or 281 day human reproductive cycle. But this is shallow indeed : it assumes that the ancient Americans did not focus primarily on the actual gestation period, but rather counted the reproductive cycle as we happen to now, namely as a 280 or 281 day period beginning with the onset of the last menstruation prior to conception and ending with the birth of the child. Consider, then, the following :

CONJECTURE : *The origin of the Tzolkin count of 260 of ancient American civilizations is to be found, in part, in attempts to model the human gestation period of approximately 267 = 260 + 7 days.*

It is well documented that the Tzolkin was originally a count of days, in fact strongly associated with the recording of birth days. Further, there are numerous possibilities whereby the gestation period could have been tabulated using the Tzolkin.

For example, consider the following natural way to execute a Tzolkin count. Build the pairs of glyphs vertically upward, always (after the obvious initialization sequence) bringing from the bottom the two next glyphs, of course one from each counter set, which form as a pair the next Tzolkin numeral. When such a Tzolkin count terminates (at 260), there is left unpaired at the bottom of the vertical column of pairs the numerals 1 through 7 of the set of 20. These 7 could then be counted to give an exact count of the average gestation period of 267 days.

Other possibilities there surely are. We mention here just one more, one which, if correct, clearly indicates an origin for the set of 13 glyphs. Assuming (brashly, I agree) that the originators became aware that days 14, 15 and 16 after onset of menstruation are on average the most likely for conception, a count to 13 was executed after the onset of menstruation, followed by fertility ceremonies including intercourse and involving a count of 7 days (using the rest of a set of 20) ; if this was followed by a complete Tzolkin count, that count would end exactly at a 267 day count after intercourse on the day of the menstrual cycle most likely for conception.

THE COUNT OF SIXTY

One of the evidenced objects of attention for early man was the periodicity of the moon. Counting single lunations, for example as 29 or 30 days, would quickly introduce a large discrepancy, and there is strong evidence[3] that Upper Paleolithic and Mesolithic man worked hard at this problem of trying to develop a workable mathematical model of the moon's periodicity, and that they succeeded in constructing a 59-day model for two lunations.

3. See A. Marshack, *The Roots of Civilization*, 1972.

Of course, many other aspects of the complex cultures of the later period centering around 5000-4000 B.C., when the notion of place-value was finally invented and sixty was chosen to be the base upon which the place-value system was erected, were involved as well, in complicated ways as yet little understood. The import of the conjecture we make here is not to deny these factors. Rather, the conjecture brings to the fore an additional factor, a new factor as yet unmentioned in the literature as far as this writer has been able to determine, that the moon calendaring of Upper Paleolithic and Mesolithic man may well have played a part.

This moon calendaring may have engendered, over the long period of its continued prominence, a considerable centrality to the number fifty-nine. Further, it should be noted in this context, the determination of a base choice of sixty because of a prominence of a fifty-nine count need not be viewed as enigmatic at all : recall that during this early period no notion of the concept of zero had yet been developed, and the building of a base of sixty upon a count of one through fifty-nine would have been as natural then as the later building of a base of ten upon the numerals one through nine. Indeed these long and steady moon calendaring investigations over thousands of years involving counts centering on fifty-nine may have led to the establishment of an importance for fifty-nine which made the later choice of sixty as the base cardinality so natural, in terms of the other factors which came into play in making sixty an advantageous choice, that the choice may have been made without any conscious awareness of the original role of the fifty-nine count.

On the other hand, it may have been that the fifty-nine day count of two lunations was done ultimately, and possibly even originally, with a count of sixty. The reason that this is a strong possibility is twofold. First, the physical phenomenon itself which was being investigated, lunar periodicity, would be most easily tracked with a sixty count since one could then count from first appearance to next first appearance. For, when observing lunar cycles, one finds the most successful practical way to do it is to count from one new moon through to the next new moon : one does not know the day to be counted as the last day of the old moon except as it is subsequently confirmed by observation of the new moon to be the day prior to such observation of the new moon. Second, such counting, "inclusive counting" as it is called, is of ancient origin, and in fact was well-entrenched even two thousand years ago, as witnessed by the *nundinae* of ancient Rome which counted an eight day periodicity of market days with a nine count which counted from one market day inclusively through the next market day. Suggesting the reader consult the first reference for further discussion of evidence, we proceed to stating the :

CONJECTURE : *The origin of the choice of sixty, as a base of ancient enumeration systems of the Near East, is to be found, in part, in very early unbased enumeration systems of the Upper Paleolithic and Mesolithic which modeled two lunations (roughly 29.5 + 29.5 days) with 59-counts and/or inclusive 60-counts.*

LOGIC : ARISTOTLE TO ARTIFICIAL INTELLIGENCE

In Evidence Logic (EL)[4], both confirmatory and refutatory predications, further annotated with evidence levels, are provided for. Thus EL's primitive expressivity exceeds that of Classical Logic. Aristotle's logic of privatives can be in part modeled in EL_2 which has only one level of evidence (absolute certainty), while the Dempster-Shafer logic for the representation of uncertainty in AI can be in part modeled in EL_n (n>1) which has n-1 levels of certainty. Provable then is the :

THEOREM : The Dempster-Shafer logic, axiomatized in EL_n, which contains p proposition symbols, k constants, and u unary predicates, has Boolean sentence algebra with ordered basis $\omega^q \cdot m^p \cdot \Sigma S_{ki} \cdot m^{ui}$ where m = n (n + 1)/2, the s_{ki} are Stirling numbers of the second kind, and $q = m^u$.

This theorem illuminates an example of current work in AI which generalizes work of Aristotle. As modeled here, the case n=2 of Dempster-Shafer logic is precisely Aristotle's logic of privatives. One sees thereby how current AI models for the handling of uncertainty like that of Dempster-Shafer, coming so long after Aristotle's preliminary grappling with logical models to address problems surrounding the concept of negation, are but one more step in the continuing process of the construction of models which help us to better understand and more efficaciously utilize our meagre, and often uncertain, knowledge about our world.

Model building is of great antiquity and is fundamental to the methodology of science. From the earliest efforts of man to understand even the simplest phenomena, we find evidenced the iteration of model construction and refinement. Further, the complex pattern of strands in this web of iteration is not only rooted deep in our past but reaches as well all the way through our long prehistory and shorter history, through the present, and (one would imagine and hope) on into an even longer future. Models are schematics of our current understanding, and are dynamic, each helping us to assess the model's faults and progress to the next refinement.

4. D. Faust, (abstract) " The concept of evidence ", *J. Symbolic Logic*, 59, (1994), 347-348 ; D. Faust, *The origin of the choice of sixty as a numeration base in the ancient Near East and pairing-generated enumeration systems of China and America*, available from the author ; D. Faust, *Evidence logic*, available from the author ; D. Faust, (abstract) " Evidence and Negation ", *B. Symbolic Logic*, 3, (1997), 364.

ANCIENT EGYPTIAN COSMOLOGICAL, ASTRONOMICAL, AND GEOMETRICAL MODELS

Kurt LOCHER

INTRODUCTION

Going back to the primitive elements of intellectual history is in some ways linked to looking at the most recent developments in modern philosophy, which are probably less ideological than those of any time since Thales. That is why the abstract by Constandache[1], prepared for another, more philosophical section of this congress, meets best my need to place Ancient Egyptian thought within a framework of appropriate notions.

Sothis as Calendar Regulator,

Ritual Metaphor,

and Pivot of the Annual Sidereal Rhythm.

For the last 150 years, the role played during three millennia by the Greek Dog Star Sirius, Egyptian Sopdet, graecized as Sôthis in the Egyptian Calendar, has been the the subject of thousands of scholarly bookpages. Recent updatings can be found in Krauss[2], von Beckerath[3], Luft[4], Clagett[5], and the present author[6]. The Egyptian civil year of 365.000 days matches Constandache's definition of a model since it is a preconception, kept during millennia despite its misfit to the yearly rhythm of nature, which follows the tropical period of 364.242 days, then most evident in the annual oscillation of the level

1. G.G. Constandache, " Paradox of Modelling in Cognitive Sciences ", *20th International Congress of History of Science, Book of Abstracts,* I, Liège, 1997, 602.

2. R. Krauss, *Sothis- und Monddaten*, Hildesheim, 1985.

3. J. von Beckerath, *Chronologie des ägyptischen Neuen Reiches*, Hildesheim, 1994.

4. U. Luft, *Die chronologische Fixierung des Aegyptischen Mittleren Reiches nach dem Tempelarchiv von Illahun*, Wien, 1992.

5. M. Clagett, *Ancient Egyptian Science*, II, Philadelphia, 1995.

6. K. Locher, " Egyptian Cosmology ", in N. Hetherington (ed.), *Encyclopedia of Cosmology*, New York, 1993, 188.

of the Nile. The difference of this tropical period from the sidereal year of 365.256 days had very probably never been noticed before Hipparch. Thus the annual periodicity of the Sothiac phenomena used to be regarded as synchronous with that of the Nile, and therefore the former considered the harbinger, the co-performer, or even the cause of the latter.

The reported[7] accuracy of prediction and observation of the heliacal rising of Sothis suggests that the discrepancy of roughly $^1/_4$ day between the mentioned periods would have repeatedly been perceived in the course of a few years, with the result that a major annual feast would slowly wander in the grid of the civil calendar. Like any star situated on the other side of the equatorial plane from the respective observer's point of view, Sothis had (and still has) a considerable period of annual invisibility for the Egyptians (and Europeans), during which it is above the horizon only at daytime.

Constandache's definition of a model, according to which the " identity of its component fades out to be replaced ", can be applied to the very case of the 70 days' invisibility of Sothis being mirrored in the 70 days' sojourn of the dead body in the embalming house before burial and transition to an assumed new life, the latter corresponding to the new visibility of Sothis after its yearly heliacal rising.

So meaningful a yearly event had caused people to look for harbingers preceding it. Since the main feature of the Egyptian (S3h)[8] Orion constellation would rise at the same point of the horizon as Sothis[9], 2 hours earlier in the same night, or 1 month earlier at the same nighttime, this most conspicuous figure in the sky was apt to announce the former event. Orion itself requiring a harbinger constellation, and so forth, a chain of correlated constellations running all around the sky westward was created, later called the belt of the decans. Indirectly this was also a motive to create spatial models (see below).

THE MODELS OF FUNCTIONALITY AND TIME

Understanding that the rising of any decan depended in a bilinear way on both the calendar date and the actual night hour must then have led to the invention of two-parametric tables, today called matrices. This occurred about 2400 B.C., earlier than anywhere in the world if we rely everywhere on the preserved documents.

Such matrices are again models, metaphorizing two-parametricity by spatial two-dimensionality, and differ from Cartesian plane representations only in

7. *Papyrus Berlin 10012 B recto*, in Luft (4 above), plate 7 b.

8. K. Locher, " New Arguments for the Celestial Location of the Decanal Belt and for the Origin of the S3h-Hieroglyph ", *VI Congresso Internazionale di Egittologia, Atti II*, Turin, 1991, 279.

9. R. Krauss, *Astronomische Konzepte und Jenseitsvorstellungen in den Pyramidentexten*, Wiesbaden.

being discrete instead of continuous. They were painted on the underside of the coffin lids of venerable persons. 17 specimens from the 22nd to the 19th centuries B.C. have survived[10]. Figure 1 shows a well preserved section containing 6 x 6 of the totally 12 x 36 matrix fields.

As in most ancient cultures, time rather than space was the predominant cosmological structuring factor. It is, however, not entirely self-evident in which sense an allegorized " time-arrow " points, the Egyptian view being partly opposed to the Greek, scholastic, and modern : " Ahead " in the meaning of " in the future " is literally rendered as " behind "[11] in pharaonic language, and the context of such sentences suggests that the idea was to go through time blindly with the back of the head towards the future and the face towards the past, seeing, *i.e.* knowing[12] it. This view survives in modern colloquial Egyptian Arabic[13], but characteristically not in colloquial Syrian[14] nor in classical Arabic[15]. Egyptians say " having something behind me " (*hâga warâ-ya*) for " having something on ".

SPACE AND GEOMETRY

As with the Babylonian Three-Path-Model[16] and with the Eudoxian celestial circles[17], geometrization of the yearly phenomena of the fixed stars also predates lunar and planetary orbital theory in Egypt. Accordingly, Eudoxian astronomy was quickly adopted in early hellenistic Egypt, from which period a pertinent manuscript[18] has been preserved that predates any autochthonous Greek counterpart (see figure 2). The decanal belt (see above), however, had been conceived long before as the probably earliest model in the world showing meridional symmetry. Accordingly, hieroglyphic writing had created symbols for the 8 cardinal and intermediate geographical directions earlier than any other scriptural system. Certainly, the almost exact south-north orientation of the flow of the Nile was helpful in developing such ideas so early.

10. K. Locher, " Middle Kingdom Astronomical Coffin Lids : Extension of the Corpus from 12 to 17 Specimens since Neugebauer and Parker ", *7th International Congress of Egyptologists*, Cambridge.

11. J. Assmann, " Zeit und Ewigkeit im alten Aegypten ", *Abhandlungen der Heidelberger Akademie der Wissenschaften, philosophisch-historische Klasse*, I (1975), 69, ref. 106.

12. German *Knowing* (*wissen*) and Greek *Having seen* (*widein*) have developed from the same indoeuropean root.

13. M. Hinds, S. Badawi, *A Dictionary of Egyptian Arabic*, Beirut, 1986, 934.

14. H. Alwâni, private communication, Damascus, 1997.

15. M. Abdel 'Aziz, private communication, Alexandria, 1997.

16. H. Hunger, D. Pingree, *MUL.APIN*, Horn, 1989, 137.

17. Hipparch, *Aratou kai Eudoxou Phainomenôn Exêgêseis*, in C. Manitius (ed.), Leipzig, 1894.

18. *The Hibeh Papyri*, I, n° 27, in B.P. Grenfell, A.S. Hunt (eds), London, 1906.

Every model dealt with so far had its more or less direct origin in sacred ritualism. In diametrical contrast, Ancient Egypt produced its planimetrical models out of the pragmatism of land surveying. Their degree of abstraction and reductionism is comparable to that of Thales some thousand years later in Asia Minor, and is best illustrated by the calculation[19] of a trapeziform area as the difference of two trigonal ones, as shown in figure 3.

FIGURES

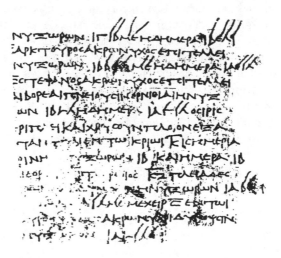

1. 6 x 6 matrix section from the astronomical lid of the coffin of Idy, 20th century B.C., Tübingen University collection.

2. Graeco-Egyptian Hibeh papyrus dealing with Eudoxian Astronomy, 3rd century B.C.

19. *The Rhind Mathematical Papyri*, A.B. Chace (ed.), Oberlin, 1927, plate 74.

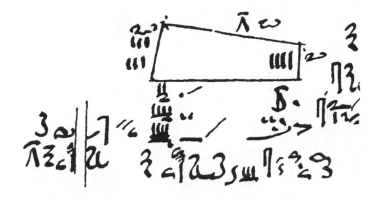

3. Calculation of a trapeziform area in the Rhind papyri, 16th century B.C., British Museum

GRAFISCHE UMSETZUNG MENTALER MODELLE : ALS PROBLEM-LÖSUNG " VORWISSENSCHAFTLICHER " TECHNIK MITTELS EXPERIMENT, GRAPHISCHEN UND RÄUMLICHEN MODELLEN[1]

Wolfgang von STROMER

Auch in den Epochen der (fälschlich sogenannten), " vorwissenschaftlichen " Technik, stellten sich unabweislich schwierigste Aufgaben. Diese verdichteten sich, seit im Hochmittelalter sich in der zunehmenden Zahl von Städten Menschen ballten. Für die anspruchsvollen Bauten der Verteidigung, des Kultus, der Herrschaft und Verwaltung, sowie der Gewerbe und für ein Stadtleben unentbehrlichen Gewässer ergaben sich neuartige Probleme. Mit beharrlichen, oft über Jahrzehnte und manchmal über Generationen durchgehaltenen Versuchen, dank Beobachtungen, Überlegungen und Erfahrungen aus *Trial and Error*, fanden sich jedoch meist praktikable und gelegentlich sogar optimale Lösungen. Dem kulturgeschichtlich Interessierten sind solche vor allem für den Hochbau der gotischen Kathedralen vertraut, wie sie seit dem Skizzenbuch von Villard de Hannecourt aus dem 13. Jahrhundert überliefert sind. Schriften über die manchmal phantastischen Entwicklungen der Kriegstechnik, seit Conrad Kysers, " Bellifortis " (Kuttenberg 1402 und Bettlern 1405), sind beim Publikum besonders geschätzt. Die hohen Ansprüche des Untertagebergbaus fanden um 1530 in den noch kaum bekannten Zeichnungen des Herrmann Groß aus den Vogesen und den Illustrationen von Manuel Deutsch für Georg Agricolas, *De Re Metallica*[2] ihren Niederschlag. Erst durch Forschungen der letzten Jahrzehnte sind die Serien der technischen Entwicklungen aufgeklärt, die eine Wasserhaltung im Untertagebergbau seit dem 14. Jahrhundert in Tie-

1. Die Stiftung Volkswagenwerk in Hannover (Az II/60 772 1) und die Hans-Frisch-Stiftung in Nürnberg (Az 33/87-14/94) förderten über Jahre mein Team-Projekt über " Brückenbau, Baustatik und Technologietransfer Italien-Deutschland/Revolution im Steinbrückenbau : Rialto-Brücke und Fleischbrücke ". Dafür schulden wir großen Dank ; denn was ich hier berichte, ist ein Resultat dieser Forschungen. Zu ihnen hat Herr Dipl. Ing. W. Brandl wesentlich beigetragen.
2. G. Agricola, *De Re Metallica*, Basel, 1556.

fen bis zu schließlich 900 Metern ermöglichten[3]. An der Entwicklung des mechanisch-halbautomatischen Drahtzugs experimentierte in Nürnberg ein Team von Draht-Fachhandwerkern, Metallurgen und Mühlwerks-Fachleuten 15 Jahre bis zur Produktionsreife, während dessen mancher Teilnehmer mangels Ertrag und angesichts der Kosten auf der Strecke blieb. Es bedurfte der Versuche und Erfahrungen von drei Generationen, bis es gelang, die Abscheidung des *ca.* 1% Silbergehalts mitteleuropäischen Rohkupfers vom Schmelztiegelchen der Alchimisten zu hochtechnischen, großkapitalistischen Schwerindustriebetrieben der Saigerhütten zu entwickeln. Je nach der Eigenart der Erze aus ihren unterschiedlichen Lagerstätten mußten, unter aufwendigen Experimenten neue und oft vielstufige Metallscheidekünste entwickelt werden. Drahtmühlen und Saigerhütten bildeten mit zahlreichen Betrieben für die folgenden Generationen in und bei Nürnberg und im Thüringer-Wald ganze protoindustrielle Reviere[4]. Übrigens regten sich gar nicht selten solche Erfindungen und Innovationen wechselseitig an, und schaukelten sich gegenseitig hoch und führten dabei zu dem besonderen, gewissermaßen archimedischen Menschentyp des Vielfach-Erfinders, der auf ganz unterschiedlichen Sektoren innovatorisch aktiv wurde. Ich versuchte jenen typischen Verlauf der sich wechselseitig anregenden Innovationen in einem Kaskadenmodell festzuhalten[5].

Solche ungewöhnlichen Anforderungen stellten zwischen 1556 und 1598 der Bau weitgespannter, möglichst flachbogiger Brücken : Der *Ponte S. Trinità* über den Arno bei Florenz, und auf kaum tragfähigem Grund der *Ponte di Rialto* über den *Canal Grande* in Venedig und die Fleischbrücke in Nürnberg. Für die Bogenführung ihrer Steingewölbe, für die Konstruktion der dafür erforderlichen Leergerüste und für ihre Fundierung gab es weder Vorbilder noch

---

3. C. Kyser aus Eichstätt, *Bellifortis*, in G. Quarg (ed.), Düsseldorf, 1967. Quarg kannte nur das (auf Rasur) König Ruprecht gewidmete Manuskript von " Castrum Mendacis " 1405 (Bettlern/Zebrak), nicht die König Wenzel gewidmete Urschrift, Kuttenberg 1402, heute in der UB Göttingen ; W.V. Stromer, " Wassersnot und Wasserskünste im Bergbau des Mittelalters und der Frühen Neuzeit ", in W. Kroker, E. Westermann (eds), *Montanwirtschaft Mitteleuropas vom 12. bis 17. Jahrhundert*, Bochum, 1984, 50-72 (Der Anschnitt Beiheft. 2) ; H. Groß, *La rouge Myne de Sainct Nicolas de la Croix*, Faksimile, Graz, 1990.

4. W.V. Stromer, " Innovation und Wachstum im Spätmittelalter : Die Erfindung der Drahtmühle als Stimmulator ", *Technikgeschichte*, 44 (1977), 89-120 ; W.V. Stromer, " Apparate und Maschinen von Metallgewerben in Mittelalter und Frühneuzeit ", *Österr. Akad. d. Wiss., phil. hist. Kl.* SB 513. Bd., Wien, 1988, 127-149 ; W.V. Stromer, " Die Saigerhütten-Industrie des Spätmittelalters ", *Technikgeschichte*, 62 (1995), 187-219 ; L. Suhling, *Der Saigerhüttenprozeß*, Stuttgart, 1976 ; W.V. Stromer, " Gewerbereviere und Protoindustrien in Spätmittelalter und Frühneuzeit ", in H. Pohl (ed.), *Gewerbe-und Industrielandschaften vom Spätmittelalter bis ins 20. Jahrhundert*, Beiheft 78 zur vswg, Stuttgart, 1986, 39-111, Karte 14, 88, *Das Nürnberger Drahtmühlenrevier*, Karte 17, 97, *Die Nürnberger, Thüringer und Tiroler Saigerhüttenreviere*.

5. Das Kaskaden-Modell von Pionier-Innovationen generell, " Gewerbereviere und Protoindustrien in Spätmittelalter und Frühneuzeit ", *op. cit.*, 90 und exemplifiziert an der " Erfindung der Drahtmühle ", *Technikgeschichte*, 44 (1977), 111-113, sowie der " Saigerhütten in Wechselwirkung mit dem mechanischen Drahtzug ", *Technikgeschichte*, 62 (1995), 213. Zum " archimedischen " Menschentyp des Vielfach-Erfinders und -Innovators vgl. W.V. Stromer, " Wassersnot und Wasserskünste im Bergbau des Mittelelters und der Frühen Neuzeit ", *op. cit.*, wie Anm. 1, mit vielen Beispielen und insbes. Anm. 67.

bautechnische Theorien. Als Ergebnis gedanklicher, dabei oft eher bauästheti-
scher Überlegungen konzipierte man gänzlich neue Kurvaturen. In Architek-
ten-Wettbewerben, mittels Ausschreiben, in Serien grafischer Entwürfe, oder
mit räumlichen Baumodellen stellte man sie den politisch Verantwortlichen
und den Bauexperten — oder in Nürnberg sogar einer, in handwerkstechni-
schen Dingen besonders erfahrenen Öffentlichkeit — zur Diskussion.

Man sorgte für Technologie-Transfer durch Anschreiben auswärtiger und
ausländischer Experten und man holte von weither Risse und Pläne anderer
neuester und auch älterer Großbauten ein. Geglückte, plausible und ästhetisch
ansprechende Lösungen fanden rasch Verbreitung. Ein Beispiel bildet der polyzentrische Korbbogen von Ammanatis *Ponte Santa Trinità* unter den Wettbewerbsentwürfen für die Rialto- und Fleischbrücke. Der *bellezza e magnifi-
cenza* maß man gleiches Gewicht bei wie der *sicurtà* und hielt anscheinend die
gelungene Baugestalt oft auch für die Lösung der baustatischen Probleme,
deren man sich bei einem Abgehen von den seit der Antike überkommenen
Halbkreisbögen durchaus bewußt war. Die Zeichnungen des Korbbogens, die
der Festungsingenieur Dionisio Boldi aus Brescia am 12. Januar 1588 als seinen zweiten Entwurf für die Rialto-Brücke einreichte, zeigt mit exakten
Details des von ihm sogenannte *Arco cavato* ein erstaunliches Verständnis für
den bei einem flachen Gewölbe enorm anwachsenden Horizontalschub. Er
schlug zu dessen Beherrschung tangentiale Überleitungen in die Widerlager
vor[6]. Kurz vor dem Bau der Rialto-Brücke wurde offenbar als Experiment und
als Modell für diese mit einer ganz neuartigen Kurvatur der Rio von Cannaregio überspannt. Mit einem polyzentrischen Bogen in Gestalt einer Parabel
führt der *Ponte delle Guglie*, fundiert auf einem Pfahlsystem im grundlosen
Fango-Morast und mit einem schräg angeschnittenen Gewölbe von 19 Meter

---

6. Der polyzentrische Korbbogen von Ammanatis *Ponte S. Trinità* ist erstmals mit Grund- und
Aufriß des mittleren der drei Gewölbe, sowie einer Darstellung der ganzen Brücke überliefert mit
Zeichnungen des Florentiner Architekten P. Cecini. Er hatte sie nebst Begleitschreiben vom 30.
Mai 1597 im Architektenwettbewerb für die Nürnberger Fleischbrücke eingereicht, auf Anfordern
des Nürnberger Ratsbaumeisters Wolf-Jacob Stromer, der sie in sein *Baumeisterbuch*, I, auf Fol.
79 und 81 aufnahm. Vgl. W.V. Stromer, ein *Lehrwerk der Urbanistik aus der Spätrenaissance*, 2
(1984) der Willibald-Pirckheimer-Gesellschaft, Nürnberg.

Ein weiterer Korbbogenentwurf aus diesem Wettbewerb aus dem Jahr 1595 anonym und ohne
Begleittext in Wolf-Jacob Stromers *Baumeisterbuch*, VIIa, prod. 41 ; nur erwähnt bei K. Pechstein, " Allerlei Visierungen und Abriß wegen der Fleischbrücken 1595 ", *Anzeiger des Germani-
schen Nationalmuseums*, Nürnberg, 1975, 72-89 ; nunmehr bei A. Bögle, *Untersuchung historischer Steinbrücken am Beispiel der Fleischbrücke in Nürnberg*, Dipl. Arbeit TU-Stuttgart, 1994,
90 Abb. 78. Die auffallend starke Rustizierung der Brücken-Strinseite rührt womöglich von einem
Florentiner Einfluß auf den anonymen Entwerfer her, seinem Stil nach vielleicht Hans Dietmair.

D. Boldi aus Brescia reichte nach einem ersten *Entwurf eines Kreissegmentbogens* vom 2.
Januar 1588 einen zweiten Beitrag am 12. Januar ein, *il disegno d'un arco cavato dal centro*.
Nachzeichnung des sehr defekten Originals bei R Cessi, A. Alberti, *Rialto, l'isola, il ponte e il
mercato*, Bologna, 1934, 197, nebst Text 366 f. u. 381 f. und G. Zorzi, *Le chiese e i ponti di
Andrea Palladio*, Vicenza, 1967, 228. Ähnlich schlug B. Lorini am 23. Dez. 1587 einen *arco
cavato dal circulo, verà fortissimo* vor, R. Cessi, A. Alberti, *Rialto, l'isola, il ponte e il mercato*,
op. cit., 351 f. Daß es sich dabei um Korbbögen handelte, wird weder von Cessi-Alberti noch von
G. Zorzi erwähnt.

Stützweite, über den breitesten Seitenkanal des *Canal Grande* nahe seiner Abzweigung[7].

Den Wettbewerb gewann indessen der Proto des Arsenals, Antonio Dal Ponte, mit Modellen für *un solo arco*. Er hatte sie — neben einem Modell für ein Drei-Bogen-Konzept — am 15. Januar 1588 und am 26. August 1588 der Kommission der *Provveditori sopra la fabrica del Ponte* di Rialto vorgestellt. Zunächst zwar hatte er sich am 20. September 1587 für die vermeintlich stabilere Dreibogenbrücke ausgesprochen. Gegen diese aber hatten sich inzwischen andere Mitbewerber mit triftigen Argumenten gewandt, die Dal Ponte alsbald übernahm. Er rechnete den *Provveditori* vor, daß die zwei Brückenpfeiler im Fangomorast der Kanalsohle nicht sicher zu fundieren wären und auch erheblich mehr kosten würden. Gesamtkosten von 42.800 Dukaten stünden nur 29.030 Dukaten für die Brücke mit nur einem einzigen Bogen gegenüber. Das waren gewichtige Argumente, zu denen die freie Passage für Schiffe und Strömung hinzukam. Entscheidend war jedoch, daß bei der Fundierung auf den seit langem verfestigten Ufern die statische Sicherheit absolut gewährt sei, und daß sie *maraveggioso*, d.h. wunderbar wirken würde. Dies sicherte Dal Ponte am 29. September und 5. Januar ausdrücklich zu. Der große Festungsingenieur Bonaiuto Lorini hatte sogar gemeint, eine Brücke mit drei Bögen samt den darauf zu erstellenden Buden wirke mickrig wie ein *pigmeo overa un nanno*, ein Pygmäe auf einem Zwerg. Er selbst hatte sich von vorne herein und wiederholt sehr nachdrücklich für nur einen Bogen mit *Grandezza und Bellezza* ausgesprochen. Am 26. August begründete Dal Ponte bei der Vorstellung seines neuen Modells, warum er ein gestuftes Fundament angelegt hatte. Auf Befragen, welche Bogenkrümmung (*circulo*) er plane, antwortete er, es fehle wenig zu einem Drittel eines Halbkreises. Nach unseren fotogrammetrischen

---

7. Der *Ponte della Guglie* überquert den großen Seitenkanal des *Canal Grande*, den Rio von Cannaregio bei der Kirche San Geremia. Die Brücke heißt nach den steinernen " Nadeln " (= *aguglie*), vier schlanken Obelisken an ihren Flanken. Aufmessung und Berechnung durch W. Brandl ergaben, daß das Gewölbe nahezu eine Parabel bildet, die polyzentrisch aus drei Kreissegmenten konstruiert ist. Das Manuskript It. VII 295 der *Biblioteca Marciana* in Venedig enthält zwischen weiteren Wettbewerbsentwürfen für die Rialto-Brücke auf carta 7r auch einen Entwurf des *Ponte delle Guglie*, der bis jüngst der Rialto-Brücke zugerechnet wurde vgl. R. Cessi, A. Alberti, *Rialto, l'isola, il ponte e il mercato, op. cit.*, 221 ; D. Calabi, P. Morachiello, *Rialto : le fabbriche e il Ponte 1514-1591*, Torino, 1987, 245, stellen dies mit Abb. 88 zwar richtig, erklären jedoc, Marchesin Marchesini für *L'autore del ponte di San Geremia*. Dies ist jedoch unwahrscheinlich, da Marchesini unter den vielen Teilnehmern am Wettbewerb für die Rialto-Brücke zu den ganz wenigen gehörte die obstinat am Drei-Bogen-Konzept festhielten und eine Einbogen-Lösung noch nicht einmal in Betracht zogen vgl. R. Cessi, A. Alberti, *Rialto, l'isola, il ponte e il mercato, op. cit.*, 361. Dagegen nahm Lorini ausdrücklich am 28. Dez. 1587 auf den *ponte di Canalregio* als Vorbild Bezug. Angesichts seiner hohen Qualitäten als Bauingenieur und des von ihm empfohlenen *arco cavato dal circulo, verà fortissimo* kommt er als Architekt jener Brücke wohl in Betracht. Deren eigentümlicher Dekor auf den Frontseiten zeigt starke Gemeinsamkeiten mit dem der Seufzer-Brücke von A. Contin, so daß auch dessen Mitwirkung anzunehmen ist. Mit einer Denkschrift vom 4. Januar 1588 nahm auch Guglielmo de Grandi, *proto et ingeniero de l'officio sopra l'aque*, zugunsten der Einbogen-Lösung Stellung. Dabei verwies er auf die beiden Brücken über den Kanal von Cannaregio. Die vor 16 Jahren von ihm restaurierte Brücke mit drei Bögen bei San Giobbe weise in den Seitengewölben starke Fundierungsschäden auf, wogegen die erst jüngst gebaute mit ihrem einzigen Bogen " sicurissioma et molto più bello " sei. R. Cessi, A. Alberti, *Rialto, l'isola, il ponte e il mercato, op. cit.*, 374.

Vermessungen bildet jedoch das Gewölbe der Rialto-Brücke einen ziemlich genauen Viertelskreisbogen von 29 Metern Stützweite und 7 Metern Pfeilhöhe. Dadurch bietet sie eine Passage auch für größere Schiffe, ist aber nur über mehr als 50 Stufen nur für Fußgänger zu überqueren. Was es mit dem angekündigten Sechstelskreis auf sich hat, ist vorläufig nicht zu klären.

Den starken Horizontalschub lenkte Dal Ponte mit einem ingenieusen Einfall in den Untergrund ab. Er setzte auf dem Ostufer, wo er ganze Komplexe überalterter Gebäude hatte abreißen lassen, um Raum für den Zugang zur Brücke und für die Fundierung zu schaffen, zwei mehrstöckige Häuser als schwere Last auf die Fundamente. Auch auf dem Westufer lastet ein Eckpfeiler der Drapperia auf den Ausläufern des Fundaments. Nur an der Nordwestflanke war das angesichts der labilen Konstruktion von *Sansovinos Palazzo degli Camerlenghi* nicht möglich. Dal Ponte suchte sich dort mit dem gestuften Fundament zu behelfen. Ausweislich einer Abrechnung vom 20. Januar 1589 wurden Dal Pontes Modelle von dem *marangon* Benetto Banelli angefertigt, der dafür u. a. vier *cartoni* verbrauchte. Der Senat gewährte am 22. Oktober 1589 Dal Ponte auf 20 Jahre ein strafbewehrtes Privileg, seine *prospettiva del ponte di Rialto fatto di pietra e della armadura fatto sotto di esso*, ein Bild der Steinbrücke und ihres Lehrgerüsts drucken zu lassen und in Venedig und Venezien zu verbreiten. Weder eines der Modelle noch ein solcher Bilddruck ließen sich bislang in Venedig wieder auffinden[8].

Nur drei und einhalb Jahre nachdem die Rialto-Brücke am 18. August 1591 fertiggestellt war, stand man in der Reichsstadt Nürnberg vor gleichen bautechnischen Problemen. Ein enormes Hochwasser mit Eisgang vom 24-28. Februar 1595 hatte sämtliche 18 Überbrückungen der Pegnitz in der Stadt und vor ihren Toren schwerst beschädigt oder sogar völlig zerstört. Die für den innerstädt-

8. R. Cessi, A. Alberti, *Rialto, l'isola, il ponte e il mercato*, op. cit., 346, 368 f., 378, 410-412, G. Zorzi, *Le chiese e i ponti di Andrea Palladio*, op. cit., 229 f. n. 55, 235, 244, n. 156, 261 zu Dal Ponte ; R. Cessi, A. Alberti, *Rialto, l'isola, il ponte e il mercato*, op. cit., 351-354, 373 f. zu B. Lorini. Einen solchen fraglichen Bilddruck stellt womöglich der bekannte Holzschnitt des *primo Modello del Signor Scamozzi* dar, der die Einbogen-Brücke als *inventione del S. Vincentio Scamozzi Architet* ausgibt, falls er binnen der Zwanzigjahresfrist erschien. G. Zorzi, *Le chiese e i ponti di Andrea Palladio*, op. cit., 249 ; R. Cessi, A. Alberti, *Rialto, l'isola, il ponte e il mercato*, op. cit., 205. Zwar berühmte sich Scamozzi in seiner *L'Idea dell'architettura universale*, Venezia, 1615, neben dem Entwurf einer Dreibogen-Brücke auch Urheber der Einbogen-Lösung gewesen zu sein, *Idea*, Parte II, Libro 8, cap. 16. Der dort für ein Libro IV angekündigte Nachweis fehlt jedoch. Libro IV liegt im Druck nicht vor und das Manuskript dafür bricht nach dem ersten Satz ab, vgl. G. Zorzi, *Le chiese e i ponti di Andrea Palladio*, op. cit., 231-235 ; R. Cessi, A. Alberti, *Rialto, l'isola, il ponte e il mercato*, op. cit., 217. Die mit demselben Holzschnitt dargestellte Dreibogen-Brücke geht allerdings auf einen Entwurf Scamozzis zurück, G. Zorzi, *Le chiese e i ponti di Andrea Palladio*, op. cit., 246 ; D. Calabi, P. Marchiello, *Rialto : le fabbriche e il Ponte 1514-1591*, op. cit., 90. Die Zuschreibung der Vogelschau-Skizze der Rialto-Brücke aus dem Wettbewerb (R. Cessi, A. Alberti, *Rialto, l'isola, il ponte e il mercato*, op. cit., 219 ; G. Zorzi, *Le chiese e i ponti di Andrea Palladio*, op. cit., 253 ; D. Calabi, P. Morachiello, *Rialto : le fabbriche e il Ponte 1514-1591*, op. cit., 92, 93) an A. Dal Ponte ist wenig plausibel. Sie entspringt eher dem Wunschdenken der modernen Autoren. Ein Kupferstich Dal Pontes von der Südfront der Rialto-Brücke aber liegt anscheinend — stark verkleinert — dem Frontisspiz von G. Franco, *Habiti d'Huomini et Donne Veneziane*, Venezia, 1610, zugrunde, G. Zorzi, *Le chiese e i ponti di Andrea Palladio*, op. cit., 263.

ischen Verkehr, wie für den Transitverkehr von sieben Fernhandelsstraßen
wichtigste Passage, die 1484 mit zwei Bögen erbaute Fleischbrücke, mußte
abgetragen und ersetzt werden. Der Ratsbaumeister Wolf-Jacob Stromer, 1588-
1614 der Höchstverantwortliche des Nürnberger Bauwesens, veranlaßte unverzüglich einen Wettbewerb mit den Nürnberger Baufachleuten und mit Experten
im In- und Ausland. Außerdem holte er Pläne und Risse bekannter Großbrük-
ken ein. Aus Venedig aber zu dem Nürnberg besonders intensive Beziehungen
pflog, kam in einem dazugehörigen Transportkasten ein zerlegbares Modell der
Rialto-Brücke und zwar genau im Maßstab 1 : 100. In halber Größe ist die
Frontseite des Modells auch auf dem Vorderdeckel des schiefergrauen Transportkastens mit Bleiweiß exakt abgemalt. Das Brückengewölbe bildet beim
Modell ebenso wie bei der Brücke selbst einen Viertelskreis — und nicht wie
Dal Pontes Modell vom 26. August 1588 einen ungefähren Sechstelskreis. Die
Geländer, Buden, Treppen zwischen ihnen und die Tempietti auf dem Scheitel
zeigen die selbe Gestalt wie auf der frühesten Darstellung der Brücke. Der
große schwäbische Architekt Heinrich Schickhardt hatte sie anno 1600, wohl
unerlaubt vom Dach seiner nahen Herberge aus gezeichnet. Dem Modell
fehlen noch die drei seitlichen Zugangstreppen und der Figurenschmuck der
Stirnseiten. Dafür aber zeigt überhaupt nur dieses Modell und sein Abbild auf
dem Transportkasten die Fundamente mit ihrer Steinschichtung, mit den Pfahl-
rosten unter ihnen. Seit der Fertigstellung des Baus sind sie ja allen Blicken
entzogen. Nur hier sieht man das fabelhafte Lehrgerüst, das sofort nach dem
Schluß des Gewölbes abgebrochen werden mußte, um die Schiffspassage wieder frei zu geben. Mit 29 Metern Spannweite, freitragend errichtet, trug es bis
zum Einsetzen der Schlußsteine im Scheitel die gesamte Steinlast des Gewölbes, ohne daß sich dessen Kreisgestalt erheblich deformierte ! Die Fundamente
an ihrer Basis und die Pfahlroste des Modells verlaufen noch waagrecht und
nicht, wie am 26. August 1588 geschildert gestuft. Den Umständen nach, han-
delt es sich also nicht um das Modell, das Dal Ponte damals vorstellte, sondern
um jenes, mit dem er am 15. Januar zunächst den Wettbewerb um den Bauauf-
trag gewann[9].

9. Zu diesem Modell und seiner Bedeutung für den Bau der Fleischbrücke, vgl. W.V. Stromer,
Lehrwerk der Urbanistik aus der Spätrenaissance, op. cit., wie Anm. 4 , 80 f. und *idem,* " Palladio
nördlich der Alpen, Nürnberg unter Wolf-Jacob Stromer ", *Bauen nach der Natur - Palladio. Die
Erben Palladios in Nordeuropa,* in J. Bracker (ed.), Ostfildern-Ruit, 1997, 170-180. mit Abb. des
Modells und seines Transportkastens, 175 Abb. 4 u. 5, demnächst *idem,* " A. Dal Pontes " Mo-
dello " ", in A. Sinopoli (ed.), *The Architecture Competition for the Construction of the Rialto
Bridge in Venice, 15th January 1588 :* Second International Arch Bridge Conference, Proceedings,
Venice, Oct. 1998 ; W. Heyd, *Die Handschriften und Zeichnungen Heinrich Schickhardts,* Stutt-
gart, 1901-1902, 39, Fig. 24, " Real Bruckh zu Venedig ", Schrägsicht von oben von der Riva del
Carbon aus, mit Maßangaben vieler Details. Die Urheberschaft am Lehrgerüst schrieb sich Dal
Ponte anscheinend selbst zu, R. Cessi, A. Alberti, *Rialto, l'isola, il ponte e il mercato, op. cit.,* 368
f., 488 ff. ; die handwerkliche Ausführung oblag Venezianer *marangoni,* die meist nur mit Vor-
namen genannt sind, Agostino, Zerbin usw. ; G. Zorzi, *Le chiese e i ponti di Andrea Palladio, op.
cit.,* 242, 259.

Modelle meist aus Holz waren in der Renaissance südlich und nördlich der Alpen als Entwürfe für Großbauten wie Kirchen, Palazzi oder Festungen weit verbreitet. Sie sind oft bezeugt und relativ häufig bis heute erhalten. Von Modellen für Brücken existiert nur noch das des Ponte di Rialto in Burg Grünsberg, das auch in der Qualität der Darstellung der bautechnischen Problemlösungen ziemlich einzigartig ist[10]. Den Umständen nach muß es auch für die Fleischbrücke erhebliche Bedeutung gehabt haben, wenngleich diese durchaus weder nach dem Vorbild dieses Modells, noch dem der Rialto-Brücke selbst, noch nach dem irgend einer sonst existierenden Brücke erstellt wurde. Gewiß gab es am großartigen *Canal Grande* und an der bescheidenen Pegnitz — die jedoch beide an ihrer Engstelle am Rialto und bei den Fleischbänken etwa 30 Meter breit waren — einige gleichartige Hauptprobleme. Beide Gewässer waren weder ablenk- noch absenkbar, beide sollten mit einem einzigen Kreissegmentbogen überspannt werden und dieser mußte in grundlosem Sumpf fundiert werden. An der Pegnitz erreichte man erst in 9-11 Metern Tiefe tragfähigen Grund. In Venedig mußte man einen Kompromiß finden für eine Passage auch für größere und höhere Schiffe und einem noch zumutbaren Fußgängerüberweg, was mit einem Viertelskreis gelang. Unter der Fleischbrücke mußte Raum für Hochwasser, Eisschollen und sonstiges Treibgut sein, vor allem jedoch eine passable Überfahrt für einen dichten Fahrverkehr der innerstädtischen Gewerbe und des Fernhandelstransits geschaffen werden, dazwischen die zum Schlachthaus getriebenen Ochsenherden und ein lebhaftes Publikum. Hinzu kam, daß die beiden Pegnitzufer sehr ungleich hoch waren und jenes zum nahegelegenen Hauptmarkt stark abfiel. Die Fronten der Brücke sollten dennoch möglichst symmetrisch wirken und sie mußte möglichst flach gespannt sein. Je flacher das Gewölbe, desto größer der Horizontalschub, der vor allem auf der Marktseite durch keine bis dahin irgendwo ausgeführte Fundamentkonstruktion aufzufangen war. Stromer suchte mittels jenes Wettbewerbs und durch Ausschreiben die notwendigen Lösungen zu finden. Dies gelang schließlich nur durch eine schrittweise Kombination von Baumodellen und grafischen Entwurfsserien, die wiederum auf mentalen Konzepten und auf deren laufender kritischer Erörterung beruhten. Diese führten zwischen den Experten der Peunt, d.h. des reichsstädtischen Bauhofs und seinem obersten Leiter dem Ratsbaumeister hin und her. Dessen Baumeisterbücher haben uns Serien von solchen Zeichnungen überliefert. Die Rialto-Brücke und ihr Modell waren als Vorbild für die Fleischbrücke ersichtlich nicht geeignet, da sie sich mit Fahrzeugen nicht überqueren ließ. Cecinis Vorschlag, nach dem Vorbild der Arno-Brücke den von ihm gelieferten Korbbogenentwurf zu bauen, hätte

10. H. Reuter, E. Berckenhagen, *Deutsche Architekturmodelle. Projekthilfe zwischen 1500 und 1900*, Berlin, 1994, 113 f. n° 269 zu meinem Modell der Rialto-Brücke. *The Renaissance from Brunelleschi to Michelangelo, the Representation of Architecture*, in H. Millon and V. Magnago Lampugnani (eds), Venice, 1994, passim ; B. Heinrich, *Am Anfang war der Balken*, München, 1979, 120-132, 87-98, *Ponte S. Trinità*, Rialto- und Fleischbrücke, letztere mit völlig falschen Maßen.

womöglich eine Lösung gebracht, kam jedoch zu spät. Die Entwurfszeichnun-
gen der sich eifrig am Wettbewerb beteiligenden Nürnberger Baufachleute
waren dagegen für eine Brücke dieser Dimension und Gestalt viel zu primi-
tiv[11].

Der Ratsbaumeister Stromer erkannte jedoch unter den 1595 eingegangene
Entwürfen an jenem des Bamberger Steinmetzmeisters Jacob Wolff d.Ä. eine
außerordentliche Begabung. Er engagierte Wolff für das Personal der Peunt
und beteiligte ihn maßgeblich an den weiteren Planungen und Experimenten
für die Brücke. Schließlich übertrug er ihm, ab Herbst 1597, die Bauleitung für
das steinerne Bauwerk. Wolff vollbrachte dort eine Rekordleistung. In knapp
neun Wochen spannte er in Tag- und Nachtarbeit über den Fluß und auf dem
von Peter Carl erfundenen und am 11. Juli 1598 fertiggestellten "Bock-
gestell" des Lehrgerüsts den Fünftelskreisbogen des Gewölbes. Es hatte 89,5
Schuh (27 m) Stützweite und 14 Schuh (4,2 m) Pfeilhöhe, gestaltet aus 3.138
erst an der Baustelle in die letzte Form gebrachten Quarzit-Keilsteinen in
mehreren Steinlagen. Schon am 15. September 1598 konnte Stromer durch das
Herausschlagen einiger Keile unter den Stützen das Lehrgerüst einlegen lassen.
Beim Sturz auf den Notsteg löste es ein Donnergetöse aus und einen Augen-
blick des Schreckens. Doch das Gewölbe hielt stand, setzte sich jedoch um
etwa einen halben Schuh (ca. 15-20 cm) in Folge der Kompression der Mörtel-
fugen. Es hält nun 4 Jahrhunderte und weiterhin, trotz eines Volltreffers einer
großen englischen Bombe im Winter 1945 auf das südliche Fundament, die
alle umliegenden Gebäude und auch die Brückengeländer einstürzen ließ, und
seither gröblicher Vernachlässigung seitens der Stadt. Der 1598 regierende Rat
aber wußte die Leistungen Stromers und seiner Mitarbeiter zu würdigen, die

---

11. Die Meinung von K. Pechstein, "Allerlei Visierungen und Abriß wegen der Fleischbrücken 1595", *op. cit.,* wie Anm. 4, 75 f., biedere Nürnberger Maurer und Zimmermeister hätten die
außerordentliche Baustatik der Fleischbrücke bewältigen können, ist mehr als naiv. P. Carl, dem
er die Urheberschaft an der Brücke zuschreibt, war zwar ein hochbegabter "Baukünstler", aber
eben doch nur Zimmermeister im Dienst der Peunt. Zwar war er der Erfinder des Pfahlrostsystems
mit den Schrägpfählen zur Aufnahme des Horizontalschubs und des hölzernen Lehrgerüsts für das
Steingewölbe, jedoch stammen weder die steinernen Fundamente mit ihrer Schichtung noch das
Gewölbe von ihm. Der Entwurf von D. Bella für die Fleischbrücke, den K. Pechstein, "Allerlei
Visierungen und Abriß wegen der Fleischbrücken 1595", *op. cit.,* 77, 84, 86 mit Abb. 12, abschät-
zig beurteilt entspricht dagegen dem Construktionsprinzip Albertis für Kreisbögen und er erscheint mir als angelehnt an den Entwurf von G. Guberni für die Rialto-Brücke. Dieser war als
Proto de i Lidi dell officio delle aque ein hochqualifizierter Experte, R. Cessi, A. Alberti, *Rialto,
l'isola, il ponte e il mercato, op. cit.,* 193 und 347 f., 397. Auch für das Lehrgerüst wurden Modelle von P. Carl gefertigt und das Gerüst selbst zunächst auf der Tiergärtnertor-Bastei aufgestellt,
um es der Kritik des Publikums auszusetzen, bevor es auf fünf Pfahlreihen samt dem Notsteg im
Flußbett errichtet wurde. Die paarweise auf den Stützen des Gerüsts angebrachten Keile
ermöglichten, die Kreisbogenkrümmung des Gewölbes nachzubessern, die sich unter der Steinlast
unvermeidlich deformierte. Schließlich erlaubten sie ein Einlegen des Lehrgerüsts, ohne daß dabei
jemand der Einsturzgefahr unter dem Brückengewölbe ausgesetzt war, vgl. die Zeichnung in
Wolf-Jacob. Stromers *Baumeisterbuch,* I, fol. 41 r, W.V. Stromer, "Palladio nördlich der Alpen,
Nürnberg unter Wolf-Jacob Stromer", *op. cit.,* 7, 177 Abb. 8 und L. Sporhan-Krempel, "Wolf-
Jacob Stromer 1561-1614, Ratsbaumeister zu Nürnberg", *Nürnberger Mitteilungen,* 51 (1962),
271-310, Abb. ab 9.

nicht nur im Gelingen des Brückenbaus bestand sondern auch in dessen enormen Eiltempo. Vor Schluß des Bogens hätte jedes der gar nicht seltenen Sommerhochwasser das gesamte Bauwerk vernichten können. Noch an der Baustelle wurde Stromer ein vergoldeter mit Dukaten gefüllter Silberpokal überreicht, in den der Goldschmied Hans Petzold vier Medaillons mit Szenen des Baubetriebs minutiös eingraviert hatte[12].

Unter der großen Zahl der Entwürfe, Pläne und Risse, die uns bis heute noch für die Fleischbrücke vorliegen, ziehen die von der Zeichnerhand Jacob Wolffs eine grandiose Spur, für die uns bisher aus der gesamten Baugeschichte kein vergleichbares Beispiel bekannt ist. In einer Serie, die von weitgehend gleichen äußeren Dimensionen des Bogens und der Fundamente des geplanten Bauwerks ausgeht, entwickelte Wolff in konsequenten Schritten zeichnerisch den Lösungsweg von der traditionellen Form — die für die angepeilte möglichst flache Bogenführung und den kaum tragfähigen morastigen Baugrund keine statische Stabilität ermöglicht hätte — schließlich die Meisterung dieser Aufgabe. Datür gab es kein Vorbild, kein Modell und keine Lehrmeinung = " Theorie ". Die Zeichnungen sind, bis auf ein paar Maßangaben, ohne jeden Text. Dennoch erlauben sie ihren Inhalt und die schrittweise Entwicklung klar zu erkennen und zu deuten. Es ging dabei um die Verbindung des Gewölbes mit den Fundamenten am Kämpfer. Wir können verfolgen, wie ein Konzept datür zuerst als Gedankenbild entsteht, dann von einer perfekten Zeichnerhand zu Papier gebracht wird, von seinem Urheber und dann von seinem Mitarbeiterkreis und dem Baumeister kritisch gewertet wird. Sodann werden unverzüglich die kritisierten Punkte und Verbesserungsvorschläge erneut durchdacht und im Gehirn zu einem neuen Bild gestaltet und das Ergebnis wiederum aufgezeichnet — und so fort, bis ein plausibler Lösungsvorschlag zeichnerisch entworfen ist, im steten Wechsel der mentalen und der grafischen Modelle. Wolffs Bilderserie beginnt mit drei aufeinanderfolgen Zeichnungen eines radial geschichteten Kreissegment-Gewölbes, das mit seiner Last auf

---

12. K. Pechstein, " Allerlei Visierungen und Abriß wegen der Fleischbrücken 1595 ", *op. cit.,* 5 und 8, 78, 81 f. Die Brücke ist 15,9 m breit. Das Gewölbe beschwert mit einer Eigenlast von 2.325 t die Fundamente und übt an den Kämpfern einen Horizontalschub von 2,60 $\mathrm{MN/m^2}$ (MN = MegaNewton) aus (Dipl. Ing. W. Brandl, *Fleischbrücke Nürnberg, Statische Berechnung,* Manuskript 1994, S. F. 29). Die Reste der zweibogigen Brücke von 1.484 wurden 1596 abgebrochen, dann wurden für die Fundierung mit eigens dafür konstruierten Maschinen 2.123 doppelmannslange Pfähle gerammt, davon 400 Schrägpfähle. Nach Grundsteinlegung am 14. Nov. 1597 setzte man das Fundament auf der Marktseite mit 5.261 Quadern in 20 Lagen, und vom 3. Mai 1598 an auf dem stabileren Gegenufer 4.659 Quader in 16 Lagen. Vom 11. Juli bis 15. September 1598 wurde das Gewölbe mit 1.872 Keilsteinen von der Marktseite und 2.166 von der Lorenzer-Seite her und zwar in mehreren Schichten übereinander aufgeführt, wie die Baustellenszenen zeigen. Diese halten in sechs Bildfolgen den Fortschritt des Bauwerks samt den Maschinen und den Arbeitern fest ; Abb. 6-9 bei W.V. Stromer, " Palladio nördlich der Alpen, Nürnberg unter Wolf-Jacob Stromer ", *op. cit.,* 176-179 und Abb. 5-9 bei Sporhan-Krempel, " Wolf-Jacob Stromer 1561-1614, Ratsbaumeister zu Nürnberg ", *op. cit.* ; W.V. Stromer, " Pegnitzbrücke Nürnberg (Fleischbrücke) ", *Steinbrücken in Deutschland,* Düsseldorf, 1988, 162-167, Abb., 164.

horizontal geschichteten Fundamenten aufsitzt. Die Spannweite des Bogens ist mit jeweils 86 Schuh angegeben.

Beim ersten Entwurf, den Wolff im Herbst 1595 unter den "Allerlei Visierungen" einreichte, ist die Pfeilhöhe 16 Schuh, dann bei den beiden folgenden Entwürfen 15 Schuh, d.h. ungefähr Viertelskreisbögen.

Der zweite Entwurf — HB 1402 — weist sonst gegenüber dem ersten nur einen scheinbar kleinen Unterschied am Kämpfer auf. Beim schließlichen Bau der Brücke kam es aber gerade zu diesem Punkt am 3. August 1598 zum Konflikt mit dem "unruhig Peter", dem Zimmermeister Peter Carl, gegen den sich Wolffs Konzept durchsetzte.

Der dritte Entwurf—HB 1427 a — bringt einen dem zweiten darin entsprechenden Vorschlag, zeichnet jedoch nicht nur ein Gewölbe von 15 Schuh Pfeilhöhe durch, sondern weit darüber eines, das dank 48 Schuh halbkreisförmig ist. Dazwischen sind nur mit Linien durchgezogene Bögen von 16 über 20, 24, 28, 32, 36, 40 bis 44 Schuh und darunter einer für 12 Schuh Pfeilhöhe.

Die folgenden drei Entwürfe hat Wolff untereinander auf einen einzigen Papierbogen gezeichnet, den Stromer in den ersten Band seiner Baumeisterbücher einfügte. Bei wiederum 16 Schuh Pfeilhöhe ist die Stützweite auf 96 Schuh gesteigert. Auf den Bögen sind Geländer skizziert. In den beiden ersten dieser Entwürfe sind die Fundamente zwar weiterhin horizontal geschichtet, die Gewölbe aber durch sie hindurch bis nahezu einem Halbkreis weitergeführt, anscheinend nach Albertis Lehre über die Statik von Segmentbögen. Dabei ist bei der zweiten Zeichnung das Gewölbe mit dem Fundament zusätzlich verzahnt, indem seine radial geschichteten Keilsteine in dieses mit derselben Schichtung jeweils eine Quaderlänge weit erstreckt werden. Schließlich bringt die dritte Skizze eine plausible Lösung für ein stabiles flaches Gewölbe, indem es durch ein durchgehend radial geschichtetes Fundament geführt wird. Das Modell der Rialto-Brücke mag dafür Anhalte geboten haben. Ein weiterer Entwurf — HB 1417 — führt dieses Konzept fort, stellt jedoch zwei halbe Gewölbe einander gegenüber mit 15 und mit 12 Schuh Pfeilhöhe[13].

Bis zum Baubeginn aber fand Stromer mit seine Mitarbeitern — Wolff, Carl, Schweher, und dazu dem Bauschreiber Friedrich Ziel/Zierl — jene Lösung, nach welcher die Fleischbrücke abweichend von Wolffs Entwürfen errichtet wurde : Ein Fünftelkreisbogen von $89^1/_2$ Schuh Stützweite und 14 Schuh Pfeilhöhe, die Fundamente zwar am Kämpfer radial zum Bogenzentrum geschichtet, in ihren weiteren Lagen aber parallel zu dieser ersten Lage und schließlich die Schrägpfähle dazu ziemlich tangential. Mit dieser originären

---

13. Germanisches National Museum Nürnberg (GNM) Hs 31700 = W.-J. Stromer, *Baumeisterbuch*, VIIa, Prod 8 ; Hs 16055 2° = *Baumeisterbuch*, VIIb (jetzt Architekturgeschichtliche Sammlung der TU München) ; dann GNM Kapsel 1061, B 1402, 1417 und 1427a ; Familienarchiv von Stromer *Baumeisterbuch*, I, fol. 31 v, Abb. 34 a-c.

Konstruktion hält die Brücke nun schon vier Jahrhundert funktionstauglich stand[14].

Die Dokumentation ihrer Entstehungsgeschichte liefert ein — bisher einzigartiges — Beispiel für einen großartigen, schöpferischen Prozeß auf dem Gebiet der Technik. Viele Umstände sprechen dafür, daß es auch bei der Planung für die Rialto-Brücke vergleichbare Vorgänge gab, wovon uns leider grafisch nur die beiden Entwürfe von Dionisio Boldi überliefert sind. Die Protokolle aus dem Wettbewerb für die Rialto-Brücke aber ergeben, wenn man sich genügend in ihre venezianische Sprache einarbeitet, einen intellektuellen Hochstand an technischem Wissen und kritischen Denkprozessen, daß man einen Begriff wie " vorwissenschaftliche Technik " als töricht und arrogant betrachten muß.

FIGURES

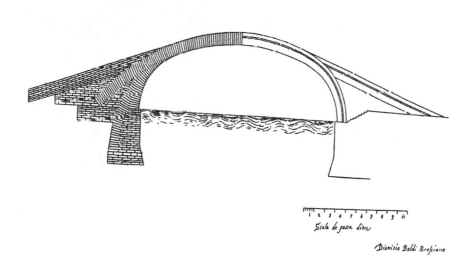

Scala de passa [...]

Dionisio Boldi Breseiano

Figure 1.

14. Vgl. die Abbildungen der fertigen Brücke mit ihren Fundamenten bei K. Pechstein, " Allerlei Visierungen und Abriß wegen der Fleischbrücken 1595 ", *op. cit.*, Anm. 9, 73 Abb. 2 ; M. Borrmann, *Historische Pfahlgründungen, Untersuchungen zur Geschichte der Fundamentierungstechnik, Materialien zur Bauforschung und Baugeschichte,* Bd. 3, Karlsruhe, 1992, 170 Abb. 164 ; A. Bögle, *Untersuchung historischer Steinbrücken am Beispiel der Fleischbrücke in Nürnberg, op. cit.,* Anm. 4, 92 f., Abb. 80 und 82 ; sämtlich nach W.-J. Stromer, *Baumeisterbuch,* I, fol. 45 v., Abb. 51. W.V. Stromer, " Pegnitzbrücke Nürnberg (Fleischbrücke) ", *op. cit.,* 165.

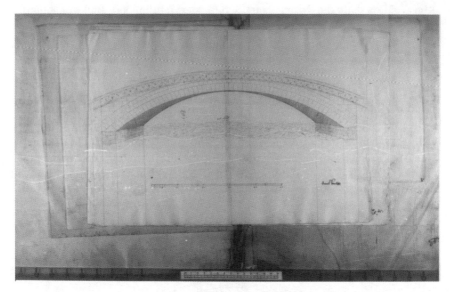

Figure 2.

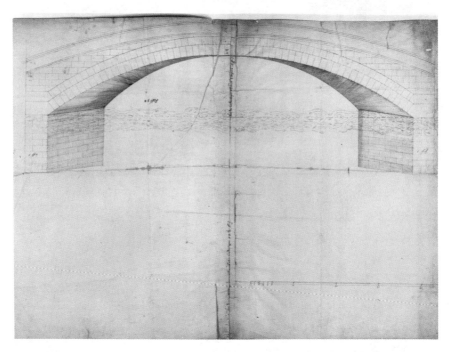

Figure 3.

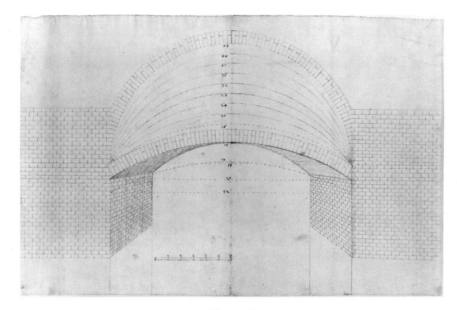

Figure 4.

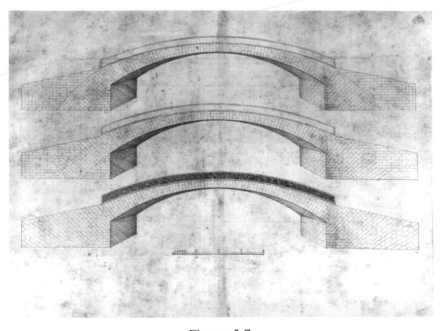

Figures 5-7.

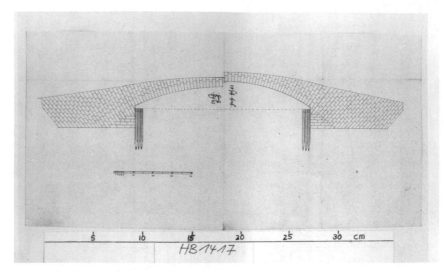

Figure 8.

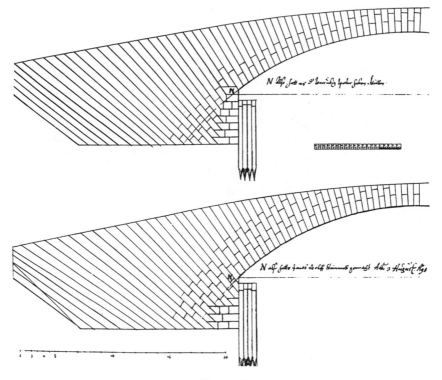

Figure 9.

BILDELEGENDEN

1. D. Boldi aus Brescia, Entwurf eines Korbbogens mit stabiler Fundierung fir die Rialto-Brücke, vgl. Anm. 4.

2-8. Sieben Entwürfe von J. Wolff d.Ä. für die Verbindung des Gewölbes der Fleischbrücke mit den Fundamenten, vgl. Anm. 11.

9. Auseinandersetzung zwischen P. Carl und J. Wolff über die Steinschichtung der Fleischbrücke am Kämpfer, Stromers Baumeisterbuch VIIb, vgl. Anm. 11.

PART TWO

SCIENCE AND PHILOSOPHY

ARISTOTLE AS THE FIRST TOPOLOGIST[1]

Peeter MÜÜRSEPP

The following analysis falls into the framework of a common tendency in contemporary philosophy of science, and probably not only of science, namely the reevaluation of the heritage of the Stagirite. Dealing with this trend, we need to pay attention not just to the Ancient and contemporary times, but also to the methodological changes initiated by some of the leading 17th century thinkers. However, our approach does not require neglecting the achievements of classical science. Aristotle is being reevaluated in a quite definite sense. The French philosopher and mathematician René Thom comments on the issue : " It seems to me that in the heart of Aristotelianism there lies a latent (and permanent) conflict between an Aristotle who is logician, rhetorician and even, when he criticizes Plato and the Ancients, sophist, and another Aristotle, who is intuitive, phenomenologist, and almost in spite of himself, topologist. It is with this second (rather misunderstood) Aristotle that I work, and I tend to forget the first "[2]. Division of the Stagirite's heritage into two quite distinct parts is not just a peculiarity of a mathematician, who is looking into Ancient philosophy hoping to find a foundation for his eclectic philosophical implications. One can find a systematical exposition of two Aristotelianisms, incompatible according to the author, also in the works of Daniel W. Graham. D. Graham speaks about the developmentalist and the unitarian view on Aristotle.

Aristotle's treatises quite likely remained in his possession throughout his life and could be modified several times. This means that it should be difficult to trace continuous development in the Stagirite's philosophy. However, there still seems to be an objective basis of dating Aristotle's certain works. It is quite possible that Aristotle progressed from Platonism to a practical empiricism. However, as we can see after a while, the opposite may hold as well.

1. My participation at the XXth International Congress of History of Science was made possible by the grant from S.A. Etilux and the travel grant from the Open Estonia Foundation.
2. R. Thom, *Semio Physics : A Sketch*, Redwood City (California), Menlo Park (California), Reading (Massachusetts), 1990, 244.

The unitarian view ignores the alleged presence of different temporal strata in the Stagirite's work. For instance, the metaphysical basis of the *Categoriae* looks quite different from that of the *Metaphysica*. Still, Aristotle himself hardly seems to recognize any historical stratification. Why should his commentators do so then ? However, the Stagirite's philosophy is not clearly self-consistent. When a historian of philosophy discovers an inconsistency, he cannot simply flag it and continue his survey.

Thus we are justified to ask, if it can be possible to reconcile the two points of view on Aristotle ? The ideal solution would probably be accommodating both views. An interesting issue rises in this connection. It has been pointed out that Aristotle develops rather towards than away from Platonism[3]. In order to establish the direction of Aristotle's development, a method involving the use of both genetic and systematic points of view should be employed. Thus we can see Aristotle as a philosopher working out the systematic implications of his methods. In this process he seems to be hammering out a philosophical system that has roots in an earlier and different position[4]. Therefore, it is possible to combine genetic and systematic points of view in an effective way.

However, there seems to be a major break in Aristotle's work which potentially serves as a fault line. According to D. Graham : " The great divide that I see in Aristotle's work is roughly coextensive with the distinction between the *Organon*, or collection of logical treatises, on the one hand and the physical-metaphysical treatises on the other "[5]. Thus, D. Graham's position is quite analogous to that of R. Thom and precedes the latter. R. Thom himself, however, claims having been unaware of D. Graham's work until his own position had been formulated.

Usually developmental interpretation has been excluded as an explanation of the great divide in Aristotle's work. In fact, there has been a general neglect of the relation of the logical to the physical works. Almost no research has been devoted specifically to this topic.

The core claim of D. Graham can be summarized as follows : " (1) There are two incompatible philosophic systems in Aristotle, namely those expressed in the *Organon* and the physical-metaphysical treatises, respectively. (2) These systems stand in a genetic relationship to one another : the latter is posterior in time and results from a transformation of the former "[6]. In the current treatise we shall not pay special attention to the relationship of the two Aristotelianisms. We would just like to stress that it was probably Aristotle the logician against whom the sharp criticism was directed in the 17th century. At the same time, hardly anyone paid attention to the inevitable result of neglecting the

---

3. D.W. Graham, *Aristotle's Two Systems*, Oxford, 1987, 10.
4. *Idem*, 11.
5. *Idem*, 15.
6. *Idem*, 15.

metaphysics of the Stagirite as well. This is just the second Aristotle according to R. Thom, who is being revived today. Quite natural reasons for such revival can be observed and are closely connected to the crisis of classical science and of the positivist-pragmatist world view.

Aristotle's philosophy is materialist in the sense that existence requires material substrate. At the same time it is teleological, being governed by form and final cause. Neglecting such association may be the main reason, why the approach of classical science has become unintelligible. Thus reviving Aristotelian metaphysics may give us a chance to restore intelligibility to our world.

Human intelligibility is inevitably ultimately dependant upon the structure of sense perception data. Vision as the most informative sense is based on recognizing significant ordered forms. The world's unintelligibility is to a large extent due to the attempt to express reality in the mathematical language. The next step would be finding the appropriate branch of mathematics. The question of form brings us to geometry. The forms of " normal " geometry are not flexible enough for an adequate description of reality. We probably need to try to understand the world in the topological manner. Following the above suggestion of R. Thom, Aristotle may help us in fulfilling this task. The Stagirite may have been looking for a synthesis between practical empiricism and theoretically precise mathematical modelling. Such synthesis is quite likely to become something like topology. Needless to say, it is not enough just to declare Aristotle a topologist in spite of himself. The validity of this claim has to be analyzed properly. This is going to be the central issue of the current treatise.

It is often useful to compare Aristotle with his master Plato in the course of analysis of any aspect of the Stagirite's thought. In the current case knowing that Aristotle did not recognize Plato's theory of ideas does not contribute to our main task to a sufficient extent. However, it is just the theory of ideas that helps us to get started. Plato has repeatedly attempted to illustrate his theory by arithmetic. This does not necessarily mean that he considered the theory of ideas to be analogous to arithmetic. However, drawing an illustrative parallel between these two theories testifies the initially arithmetical attitude of Plato, in spite of the legendary writing over the gate of the *Academia*. Therefore, we can say that Plato took the stand of the purest *a priori* science according to Henri Poincare, which happened to be the wrong choice. Nevertheless, it is still geometry that occupies the centre of H. Poincare's hierarchy of sciences.

First, we have to show, if the Stagirite was more a geometrician than his master. Geometrical thinking implies possessing at least intuitive understanding of the notion of continuity. Plato's world view was strongly influenced by Pythagorism. This becomes clear first of all in the *Timaeus* where Plato makes the Demiurge to construct the polygons and polyhedrons which constitute the elements. Plato was aware that the moving point generates a curve, a moving curve generates a surface and a moving surface generates a volume. It seems that Plato has interpreted these generations as of the type of discrete generativity. If so, the idea of polygons becomes natural. The point as a pure zero could not serve as a base for the construction. It needed to be "thickened". This kind of thinking shows that Plato probably did not pay continuity any special attention. It may be that Aristotle, on the contrary, held number generativity in disregard. R. Thom claims that the Stagirite did think about the world in terms of the continuous. By the words of R. Thom, Aristotle's revolt against Plato is that of a topologist against arithmetic imperialism, that of the apostle of the qualitative against the quantitative[7]. This kind of emotionally powerful rhetoric, however, gains no meaning without thorough rational analysis.

The problem of Aristotle's disagreement with Plato is closely linked to the question of the relationship between mathematics and reality. It is difficult to pose this problem in any other way than philosophically. Normally, its essence is the relation between continuous and discrete. Plato held that the best intelligibility of reality can be achieved by modelling it in the discrete manner. Absolutely adequate reflection was not accessible anyway as material reality was imprecise in itself. Aristotle, however, distanced himself from such approach by trying to achieve true intelligibility of reality. The Stagirite did not view the world against any kind of theoretical background. He did not attempt to fit reality into a Kantian type schema. Aristotle's ideal was taking the world as it is and describing it in an abstract manner. Therefore, his vision of the material world, presented mostly in the *Physica*, is based not on numbers but on continuity.

The Stagirite's achievement is truly remarkable in the context of R. Thom's works, who has dreamed of developing a Mathematics of the continuous[8]. Such task cannot be purely mathematical but necessarily implies a philosophical program. By R. Thom's own words this program involved the geometrization of thought and linguistic activity[9]. With the help of catastrophe theory he has attempted to overcome the classical problem of the relationship between continuous reality and discrete mathematics by creating a new kind of mathematical method, enabling to formalize abrupt changes in reality. R. Thom openly takes up natural philosophy, trying to build up an intelligible ontology based upon Aristotelian philosophy of nature.

Now we can further concretize our aim. It lays in the evaluation of R. Thom's analysis of the hypothetical primordial topological intuitions of Aristotelianism. We shall start with the bold claim of R. Thom that Aristotle basically postulates the notion of continuity having multiplied continua to the point of making the continuum the single substrate of all mobility[10].

---

7. R. Thom, *Semio Physics : A Sketch*, op. cit., 166.
8. *Idem*, viii.
9. *Idem*, viii.
10. *Idem*, 166.

Aristotle did not recognize space in the Cartesian sense. His metaphysics required space to be the predicate of the substance (the *topos*). The opposite does not hold. It is true that each type of change (*metabole*), each genus, requires a specific matter for the Stagirite. But all these matters have one thing in common : they all have the character of spatial extension. Therefore, they are all continua. Aristotelian space is never empty. It is the place for entities (*ousia*). Such approach to space makes the Stagirite multiply the kinds of matter developing thus the basis for continuity. Aristotle defined prime matter as the essential subject of all opposition of contraries. Thus, matter ought to be a " subject " able to take the predicates " extended and not extended " indifferently. But the Stagirite rejected the predicate " extended " alltogether. As a result all matter becomes endowed with continuity in an undifferentiated sense[11]. Thus, Aristotle has managed to equalize matter with qualitative space.

Aristotle's approach is definitely different from Plato's dualism, but does not make him necessarily a topologist. We have to become more specific in order to form our final opinion about the issue. Careful reading of the *Physica* reveals interesting aspects of the Stagirite's thought. It is quite clear that Aristotle has made a difference between the character of natural numbers and continuous magnitudes. One can just express astonishment that so little notice has been taken of this fact before it was pointed out by R. Thom. In the *Physica* does the Stagirite observe that from the subtraction point of view, continuous magnitudes and natural numbers have contrasting behaviour. Any integer subjected to an unlimited sequence of subtraction of one is eventually exhausted and annihilated. A continuous magnitude, however, can suffer an infinity of subtractions of continuous magnitudes without vanishing[12]. By adding the dimensions subtracted from the original magnitude to what is left, in the same temporal order, one builds a sequence of increasing magnitudes which admits a bounded limit[13]. The space so constructed never reaches its limit. In the modern sense it is an open set. The Stagirite writes : " ...but being whole and infinite not in itself but in respect of something else. Nor does it surround, *qua infinite*, but is surrounded... "[14]. Aristotle seems to be speaking about a bounded open set. A few pages below one can read : " ...for the extremes of what surrounds and of what is surrounded are in the same (spot). They are both limits, but not of the same thing : the form is a limit of the object, and the place of the surrounding body "[15]. Aristotle seems to have considered the thickness of the extremes to be negligible. If so, he must really have been very close to modern mathematical thinking.

11. R. Thom, *Semio Physics : A Sketch, op. cit.*, 166-167.
12. Aristotle, *Physica*, III, 207b.
13. R. Thom, *Semio Physics : A Sketch, op. cit.*, 167.
14. Aristotle, *Physica*, III, *op. cit.*, 207a, 24-25.
15. *Idem,* 211b, 12-14.

The Stagirite has expressed the position that the infinite has an intrinsic sub-strate, the sensible continuum[16]. This fragment brings us right to the heart of the matter. A species (*eidos*) has no substrate proper to it, unless one associates to it its own " extension " in the space of the genus. If we consider the distinction finite-infinite as sub-tending a genus, we may think that the finite has number for its intrinsic substrate, while the infinite has the continuum[17].

It is an important point to remember that, for Aristotle, everything pertaining to numbers is in possession of matter, even the infinite[18]. This enables R. Thom to come to a debatable conclusion that there is no pure discrete number for the Stagirite. It may be so in a specific context. However, the context may really be in place, as far as materiality is concerned. It is justified to say that every discrete being is realized by a continuous figure[19]. This may be the Stagirite's solution of the Pythagorean problem of numerical modelling.

We can get even closer to the primordial topological intuitions of the Stagirite, if we look into his conception of the " real " world. Let us pay attention to another quote from the *Metaphysica* : " For the incomplete spatial magnitude is in order of generation prior, but in order of substance posterior, as the lifeless is to the living (…) If, then, that which is posterior in the order of generation is prior in the order of substantiality, the solid will be prior to the plane and the line. And in *this* way also it is both more complete and more whole, because it can become animate. How, on the other hand, could a line or plane be animate ? "[20]. R. Thom offers a topologist's interpretation of the quote. He considers it likely that closed balls in space are the only " real " entities for Aristotle. An animal has a body that is a ball in space. The boundary of the animal, its skin, is a part (*meros*) of the animal. Next comes the question, if the complementary to the skin, the inside of the animal, can be considered as a part too. This question brings us to the very interesting problem of the so-called homoeomerous parts.

The initial idea of homoeomerous and anhomoeomerous parts probably belongs to Anaxagoras. The notion of the homoeomerous itself, however, can be purely Aristotelian. The notion is tied to spatiality and infinity at the same time, homoeomerous being an entity that can be divided infinitely without ceasing to be itself. Cutting a piece of gold into halves always yields two pieces of gold, whereas the division of a living organism (at least in the case of vertebrates and birds) never permits to obtain two individuals of the same kind. The Stagirite seemed not to be aware of the limits of divisibility. The homoeomerous is a controversial issue in Aristotle's philosophy. In fact, it is

16. Aristotle, *Physica*, III, *op. cit.,* 208a.
17. R. Thom, *Semio Physics : A Sketch*, *op. cit.,* 168.
18. Aristotle, *Metaphysica*, Ë, (8), 1074a.
19. R. Thom, *Semio Physics : A Sketch*, *op. cit.,* 168.
20. Aristotle, *Metaphysica* M, 1077a, 19-20, 25-31.

matter without a definite form, still not being the prime matter. As we shall see below, homoeomerous parts should be taken as potentiality which is never in act. A quite strange position.

There are certain qualities which require spatiality, like colour. Corporal primary entities like man and animals can be simultaneously predicated according to contrary species. For instance, an animal may be black and white. Here the conjunction obviously does not have Boolean signification. In this way one can define the parts of the whole, assuming that for each part a predication has a Boolean character. Thus the skin of a black and white animal will be separated into a black part and a white part. R. Thom sees here the origin of the notion of homoeomerous parts[21]. Connecting such argumentation with Aristotelian thinking may well be an exaggeration of a contemporary mathematician. However, if we consider the distinction of prime matter and local matter by the Stagirite, R. Thom may turn to be right.

Still, we need to take a closer look at the situation of opposite predication. What does it mean that one subject admits two opposites as simultaneous predicates ? In the case of black and white, it just means that the subject is not only white but also black at the same time, demonstrating the subject's extended character, its spatiality. In such case no contradiction rises. A subject of extended character may freely admit opposite predicates simultaneously. (As far as colour is concerned, the subject certainly always has spatial character). This point was left unnoticed by the Eleatics. Being need not be eternal and indivisible as it is extended. Purely logical contradictions do not matter in the sensible world. Aristotle demonstrated remarkable consistency of thought in keeping this distinction in mind.

So, if we suppose the inside of a living being to be homoeomerous, then the inside is an open ball representing the state of potentiality, *matter*, whereas the boundary sphere represents form[22]. In such manner one gets the chance to interpret material reality in topological language which is free from the limitations of the formalisms of classical mathematics.

However, one should be careful at this point. The last claim of R. Thom may seem just quite primitive natural philosophy. On the one hand, he has never denied that it is just this field of knowledge he is attempting to revive. On the other hand, analysis of the basic Aristotelian conception of matter and form do really reveal astonishing parallels with R. Thom's own thinking while working on catastrophe theory. Obviously, in the case of solid bodies form and matter coincide, form being the permanent origin of the quiddity. " The form of the boundary of a solid is the mark of the catastrophe that separated this solid from a pre-existing material, or perhaps of the particular circumstances

21. R. Thom, *Semio Physics : A Sketch*, op. cit., 167.
22. *Idem*, 169.

surrounding its genesis : solidification or precipitation of a saturated solution "[23].

However, we need not remain on the level of philosophy of nature. One can interpret the transformation of potentiality into actuality in the topological manner as well. For instance, the classical definition of motion as *the entelechy of the potential being as such* can be interpreted according to the axiom : " The act is boundary to the potential ". For the purpose of brevity we denote this axiom as ABP. The latter is analogous to another axiom : " Form is boundary to matter "[24]. This analogy enables us to move further, saying that just as form is the boundary of the entity whose inside is the material support, the definition is the boundary of a notion in the space of " intelligible matter ". This is in perfect accord with a central principle of the Aristotelian system : individuality based on separation. The principle is at work in both material and intellectual world.

The axiom ABP has interesting implications in biology. One may attempt to present a more precise interpretation of the problem of homoeomerous parts. At this point R. Thom sees a direct link with his notorious catastrophe theory. The theory includes the notion of regular point or a point of regular structure. But we have to consider homoeomerous parts of dimension smaller than the maximum, *i.e.*, three in the case of Euclidean space. Dealing with the body of a living being, it is possible to describe its division geometrically. To say in other words, stratification becomes possible. Homoeomerous parts happen to be strata as well. Blood, flesh, the interior of bone are all homoeomerous parts. The essence of anhomoeomerous parts, however, is not that clear. They are reunions of homoeomerous parts. But their global principle of individuation is said to be lacking[25]. Anhomoeomerous parts can be made up of homoeomerous parts. The opposite can hold by no means.

An interesting issue rises in connection with the axiom ABP. It appears to be in patent contradiction with the catastrophic interpretation of the hylomorphic schema[26]. Potentiality suddenly happens to be boundary to the act. It is not necessary here to follow R. Thom's sophisticated formalism. Instead, we are going to look into the Aristotelian origin of the issue, which is connected to the question of a becalmed fleet, of actual immobility. A fleet moving before the wind seems to be in stable motion. However, such fleet cannot be in act from the Aristotelian viewpoint, as the ships are constantly changing place. In the case of the fleet there is no evident canalization or separation. But, in fact, the fleet's motion is not unlimited. An enveloping form of the motion is represented by the coastline. Therefore, possible states are clearly separated from

23. R. Thom, *Semio Physics : A Sketch, op. cit.*, 171.
24. *Idem*, 169.
25. *Idem*, 171.
26. *Idem*, 175.

impossible ones, although the area of possible states may be very large. Anyway, there does exist a certain kind of canalization. Breaking down of this canalization leads to the hylomorphic schema of catastrophe theory. Privation in the form is necessary for entry into the metastable state. In the case of the fleet a real privation can hardly be imagined. But there certainly exist real situations, where privation does occur. For instance, when a watchdog breaks his chain.

Privation is a necessary preliminary to the transition act-potency. But what is privation ? Is it just the disappearance of form ? Certainly not. It is, in fact, a certain type of form, a form with a gap in it, as a hollow in the boundary sphere. Aristotle himself calls privation a form of sorts or species[27]. The transition into potency initiated by the privation may be followed by an act " sorted out " by the privation and fulfilling it. Thus, this act also realizes the potential.

How, in the end, are we to distinguish between act and potency ? On the one hand, the describable act should be seen as morphologically fixed and so characterized mathematically by a constant state value[28]. In this case, the state of potentiality would be characterized by an intrinsic variability. On the other hand, an inverse symbolism can also be allowed, as the organizing centre includes the unfolding which it generates. Potentiality would be located at the organizing centre. Thus, actualization would be an unfolding of the unstable situation.

The main difficulty can be formulated in the Aristotelian manner. In biology the difficulty can be in attributing a relative potential or actual character to an anatomical morphology. It is particularly visible in the case of a joint like that of the elbow between proximal bone (*humerus*) and distal bone (*radius*)[29]. The outstretched arm can be considered as one in direction, the elbow-joint exists only in potentiality. The joint becomes actual, when it is flexed. " Considered in the space of directions, the stretched position where humerus and radius are the prolongation of each other along the same line is a situation algebraically degenerate by comparison with the bent position where the directions are distinct. So Aristotle regards the algebraically degenerate situation as in a state of potentiality, the generic situation as actual "[30]. The described situation happens to be the opposite one to the ABP axiom. The state of potency (the stretched position) is boundary to actual situations. Thus, it is hard to associate an intrinsic interpretation with a given geometrical situation. The interpretation is necessarily dependent on the biological function of the organ in question.

In the above sense, Aristotle's biology is still closer to natural philosophy than to topology. However, the Stagirite demonstrates an algebraic vision by

27. Aristotle, *Physica*, II, 1, 193b, 19.
28. R. Thom, *Semio Physics : A Sketch, op. cit.*, 176.
29. *Idem*, 177.
30. *Idem*, 178.

placing the soul (the *arche* of motion) in the organizing centre at the most degenerate point. Consider the following : " And the middle of the body must needs be in potency one but in action more than one "[31]. Thus, there is an actual plurality of a potential unity.

The latter is closely connected to the formation of the conception of quiddity. In the case of the permanent form the origin of the quiddity is linked to the permanence of a substrate. One can see that the connection has been established in the manner of natural philosophy. However, metaphysical character is not totally lacking. A return to the organizing centre is quite conceivable. It can be demonstrated by the cusp model of catastrophe theory.

One can take a closer look at the problem of quiddities and genera. Homoeomerous parts have no form in the spatial sense. Still, Aristotle attributes a *logos* to them. How to understand this ? In fact, the Stagirite may have been ambiguous in his attitude towards the divisibility of matter. The infinite hardly exists in actuality. Therefore, one should expect Aristotle to refuse an infinite divisibility of the continuum. If the latter would be the case, continuum would become actual. However, in the fourth book of the *Physica* the Stagirite claims that " There is no continuum without parts "[32]. What about qualities like dense, rough, polished ? Aristotle seems to admit that such qualities are due to an arrangement of very fine parts, at the limits of the visible. Now the question rises, how to associate a qualitatively defined genus with this invisible order ? R. Thom writes about the issue : " We moderns know (in theory) how to define temperature by averaging the speeds of ambient molecules. But this operation is one of the most mysterious acquisitions of modern science "[33]. Ludwig Boltzmann's irreversibility comes from reversible Hamiltonian dynamics of discrete particles. We cannot deal with L. Boltzmann thoroughly here. Let us just stress that passing the thermodynamical limit can be interpreted as constituting a continuum. The thermodynamical limit is usually defined by dilating to infinity a box containing a gas of hard balls. It can also be defined by keeping the box fixed and dividing it up into finer and finer cells. However, it is very probable that there exists a limit to such operation. For instance, anything concerned with life has an undivisible part, the cell. Therefore, R. Thom is very much right to state that the problem of quiddity is still with us[34].

The Aristotelian problem of quiddity may be the very key issue for understanding the world. However, dealing with the issue leads us to the problem of evaluation of the whole heritage of the Stagirite. Do we have to study Aristotle ? This question probably has to be analyzed separately from the point of view of a philosopher, a mathematician and a scientist. The above-men-

31. Aristotle, *Metaphysica, op. cit.,* 702b, 26.
32. Aristotle, *Physica, op. cit.,* 233b.
33. R. Thom, *Semio Physics : A Sketch, op. cit.,* 186.
34. *Idem,* 186.

tioned L. Boltzmann, for instance, approaches the problem from the scientist's point of view : " Qualitative Aristotelian science does not have to be taught because part of it is naturally assumed by virtually everyone anyhow, much of it only holds under very special circumstances based on false understanding, and also because in a number of important circumstances it doesn't hold at all... "[35]. Different scientific theories describe certain fragments of the world to a certain extent of precision. The greater the precision, the better the applicability of the theory. However, a modern scientist often does not care about understanding the world or even understanding his/her theories. The only thing that matters is applicability, either practical or theoretical. The Stagirite's attitude was never like this. His philosophizing constituted a continuous attempt to think himself into the world, to understand it. Aristotle's intellectual activity was not harassed by the constant pressure to innovate the *Umwelt*. He was free of the modern man's strife to control the world by placing himself out of its boundaries. The Stagirite considered himself a legitimate member of the sublunar world.

Fortunately, the question of teaching Aristotelian mathematics did not come up in the past and it hardly could. The Stagirite seemed not to have made any remarkable direct contribution to mathematics. Such opinion was very natural as the impressions of mathematics being qualitative at some point remained nonexistent or vague. Only quite lately it became possible to think that, in fact, there exists a very interesting approach to mathematics in the works of Aristotle. Qualitative mathematics of the Stagirite was discovered just in time. Today it receives enough attention in the light of the crisis of positivist-pragmatist approach to the world. Interest to qualitative mathematics rises alongside with qualitative approach to natural science. It has become more and more clear that, contrary to L. Boltzmann's opinion, it may be just qualitative physics of Aristotle that we need to follow in order to make our world intelligible. Such opinion is nothing original having been put down already by L. Boltzmann's contemporary Pierre Duhem. P. Duhem notices some analogy between physics of his day, namely thermodynamics, and the Aristotelian cosmology reduced to its essential affirmations[36]. P. Duhem claims to recognize in Aristotelian cosmology and thermodynamics two pictures of the same ontological order[37]. Thermodynamics with its second principle certainly represents a step away from classical Newtonian science towards a novel qualitative approach developed by contemporary natural science. In the light of the above analogy it is natural to look for the same kind of connections and developments in the field of mathematics. It is may-be to early to ensure that we have succeeded easily

35. " A Documentary History ", *Ludwig Boltzmann. His Later Life and philosophy, 1900-1906*, Book I, Dordrecht, Boston, London, 1995, 23.
36. P. Duhem, *The Aim and Structure of Physical Theory*, Princeton, New Jersey, 1954, 310.
37. *Idem,* 310.

just following the path of R. Thom. However, we can be quite confident that we have been looking in a very promising direction.

Some primordial topological intuitions can definitely be found in Aristotle. Success of the search, however, strongly depends on the aims of the interpreter. The notion *syneches*, for instance, plays an important part in the Stagirite's philosophy. Still, its semantic field in the Ancient times was too broad and vague for constructing a prototype of the modern term of continuity from it. In fact, the Aristotelian notion covers three modern mathematical notions : continuity, connectedness and the monotony of function.

Needless to say, Aristotle was not a topologist in the modern sense of the term. However, a modern scientist or mathematician active in whatever field can only gain by studying the creations of the amazingly powerful and flexible mind of the Stagirite. At the same time, he has constantly to keep himself aware of the serious danger of overinterpreting Aristotle's writings due to the wide gap in space-time between the Stagirite and the contemporary thinker, which can never be bridged in a proper way.

SCIENCE AND PHILOSOPHY IN THE ANTHROPOMETRIC STAGE

Bert MOSSELMANS and Ernest MATHIJS

INTRODUCTION

This paper investigates hidden premises of science in classical Greek antiquity. We describe the " anthropometric stage " or the " anthropometric world-view " as an unconscious framework on which society, thought and science are based. The " anthropometric stage " does not refer to a concrete process of historical development, but it is a statical description of an unconscious world-view. The anthropometric world-view is present in classical Greek antiquity, and becomes superseded by the lineamentric world-view during the Middle Ages[1].

Kula and Mirowski identify three different stages in western history : the anthropometric, the lineamentric and the syndetic stage[2]. According to Kula, " old " measures cannot simply be translated into modern metric equivalents, because they contain a hidden social content. Whereas in the " anthropometric stage " metrological concepts correspond to parts of the human body, in a later stage reference is made to " units of measure derived from the conditions, objectives and outcomes of human labor ". Kula emphasizes the role of measurement as an attribute of authority. Mirowski extends Kula's conception of the metaphors of measurement to the central metaphors of motion (physics), of the body (anthropomorphics), and of value (economics). In the anthropometric stage these metaphors are unified ; they become separated during the lineamentric stage ; and they are finally re-unified in the metaphor of " energy " during the syndetic stage. Whereas Kula stresses the hidden social content, the social aspects are minimized in Mirowski. In order to re-introduce social conditions

---

1. O. Löfgren, " World-Views : A Research Perspective ", *Ethnologia Scandinavia*, 11 (1981), 21-36.
2. W. Kula, *Measures and Man*, Princeton, 1986 ; P. Mirowski, *More Heat than Light*, Cambridge, 1989.

we refer to the theory of mimetical desires described by René Girard[3]. The " stages " are differentiated from each other by the regulation mechanism which accounts for a channeling of mimetical desires. In the anthropometric stage the " scape-goat mechanism " defines a society based on the value of " human inequality ". This society is compatible with " anthropometric " measurement and the principle of self-determination.

GIRARD AND THE SCAPE-GOAT MECHANISM

Girard starts from an anthropological point of view — he focuses on differences between human and animal societies. He devotes special attention to " mimesis ", the acts of imitation directed solely towards the possession of certain goods, material as well as immaterial in nature. In the animal world this mimesis concerns only a dominant figure possessing wanted goods, like food or females. Animals subject always to this dominant figure through a certain ritual which is not truly violent in nature. This subjection does not appear in the human world, where the struggle for material objects and prestige shows a tendency to degenerate into violence[4].

The mimesis directed towards possession can be tempered only when it turns into a conflict mimesis. In this, two or more persons unite themselves and direct all violence towards one person, the scape-goat. When we look at the struggle it is difficult to see why this person is chosen, because the struggle takes place between equals. However, inside the group the choice is clear, because the scape-goat becomes differentiated from the rest as the cause of the encompassing violence. In Girard's analysis of reports of collective acts of violence, some stereo-types are identified : a decline of differences, as a general description of the social and cultural crisis ; some offenses which are thought to be connected with the crisis ; and the identification of a guilty person, which has characteristics differentiating him from the rest of the group, but these characteristics have nothing to do with the crisis in itself[5].

The scape-goat contains two important aspects : he is *maléfique* (" bad "), because he represents the violent crisis ; but he is *bénifique* (" good ") as well, because the community becomes re-unified through its coalition against him. The killing of the scape-goat has the effect of a miraculous liberation, because it marks the end of a horrifying mimetical process. Everything will be done to avoid similar situations in the future. The *maléfique* aspect gives rise to the

3. R. Girard, *La violence et le sacré*, Paris, 1972 ; R. Girard, *Des choses cachées depuis la fondation du monde*, Paris, 1978 ; R. Girard, *Le bouc émissaire*, Paris, 1982.

4. R. Girard, *Des choses cachées depuis la fondation du monde, op. cit.*, ch. I.I.A, I.III.B, III.I ; R. Girard, *La violence et le sacré, op. cit.*, ch. VI.

5. R. Girard, *Des choses cachées depuis la fondation du monde, op. cit.*, ch. I.I.D. Girard's analysis of the Oedipus myth provides a good example. See R. Girard, *Le bouc émissaire, op. cit.*, ch. III.

establishment of several prohibitions, in which mimetical desires directed towards possession will be put under restraint. The *bénifique* aspect leads to a recurring need to remember the miraculous liberation in the ritual. In the ritual humans or animals representing the original scape-goat are sacrificed in circumstances as similar as possible[6]. It is important that this scape-goat mechanism remains unconscious : as soon as some members of the community " discover " the arbitrariness of the chosen victim, the mechanism will no longer work[7].

ANTHROPOMETRIC MEASUREMENT AND THE PRINCIPLE OF SELF-DETERMINATION

In order to temper mimetical desires, the differences between humans are stressed and institutionalized : unequal humans are less inclined to compare themselves with each other. Traditional people recognized the role of inequality and hierarchical relations in the temptation of mimetical desire and violence[8]. Achterhuis argues that in traditional societies *hierarchy* formed the predominant value, in contrast to the value of *equality* in contemporary western civilization. This hierarchy is closely tied to the existence of a solid societal order, in which every entity performs its proper function[9].

The existence of inequality as the predominant value gives rise to the establishment of a *hierarchically differentiated* social order, in which every entity performs its proper function. This signifies that there exists a hierarchical societal pyramid, but also that a person higher on the pyramid is not necessarily permitted to perform all actions which " lower " persons are allowed to do.

In a hierarchically diffcrentiated social order with human inequality as foundational value, there is no need for abstract principles of measurement. Measures are an attribute of authority. In classical Greek antiquity measures, like coinage, were an attribute of the sovereign power ; newly emergent city-states used proper measures to symbolize their sovereignty, whereas conquered places had the measures of the conqueror imposed on them ; and in certain instances, the king used weights larger than the normal ones[10]. In the anthropometric stage, man measures the world by himself, by employing his body parts (foot, arm, finger, etc.). Since men are naturally regarded to be unequal, it is only natural to relate measures to individual, concrete persons, and to their social status. Protagoras' statement that " man is the measure of all things " is therefore simply an expression of the existing state of affairs[11]. The existence

---

6. R. Girard, *Des choses cachées depuis la fondation du monde, op. cit.,* ch. I.I.D.
7. According to R. Girard, the judeo-christian tradition discovers the scape-goat mechanism.
8. R. Girard, *Des choses cachées depuis la fondation du monde, op. cit.,* ch. I.IV.A.
9. H. Achterhuis, *Het rijk van de schaarste,* Utrecht, 1988, 121.
10. W. Kula, *Measures and Man, op. cit.,* 18.
11. W. Kula, *Measures and Man, op. cit.,* 24-25.

of abstract measures would imply that criteria applying to one situation could be transmitted to other circumstances. Since the societal positions of the individuals have to be taken into account, this transmission does not take place. This implies that " abstract " measures are irrelevant in the anthropometric stage.

Compatible with the fact that every individual measures the world around him " by himself ", the predominant principle in Greek thought states that every entity is explained " by itself ". Whereas measures remain closely tied to the persons and objects involved in the process of measurement, the explanation principle of self-determination implies that every entity acts according to a certain essence which is in accordance with these objects. Like the measures, these essences are not related to each other by any set of fixed conversion factors : the essences are mutually incommensurable. The emergence of Greek thought should be placed within the context of Girard's description of ritual and myth. The ritual tries to remember the killing of the scape-goat in circumstances as similar as possible to the original act of violence. In mythology the original violence is hidden : the three stereo-types already mentioned above do not report the arbitrariness of the violent act, the scape-goat is presented as guilty by nature[12]. Whereas in mythology forces are personified, in early Greek philosophy an abstract mean appears, removing the direct contact between man and world[13]. Early thinkers as Thales sought explanations within the boundaries of a rational and intelligible order, concealed by the visible world[14]. Greek thought in fact tried to build a " rational " mirror of the surrounding world, by identifying the " natures " or " essences " of different entities, thereby incorporating *unconsciously* the general principles on which the society was based. Therefore philosophy represents a further abandonment of the original (arbitrary) violence, because the world *as it is* presented as conforming to a natural order. Since anthropometric society is based upon hierarchical differentiation, the " rational mirror " of this world will have the same features[15].

We conclude that the *anthropometric stage* or *world-view* contains :

1. hierarchical *differentiation* in the social order ;

2. *anthropometric measurement*, which means that every individual maintains his own measures, according to the parts of his body and referring to his social position ; and

12. R. Girard, *Des choses cachées depuis la fondation du monde, op. cit.,* ch. I.IV.A.

13. H. De Ley, *Geschiedenis van de wijsbegeerte der Oudheid,* Brussel, Gent, 1983, 6-8.

14. W.K.C. Guthrie, *A History of Greek Philosophy,* I, Cambridge, 1962, 29.

15. We do not defend a " materialistic " or an " idealistic " view, in which ideas are depicted as derived from material conditions or in which the material world is seen as a collection of objectified ideas. The ideas concerning " inequality " and " the natural order " are part of an unconscious world-view, which is expressed in thoughts, actions and social structure.

3. the explanation principle of *self-determination*, which states that every entity acts according to its own essence. This essence can be placed within a hierarchical system, but without simple mathematical relations or fixed conversion factors to other essences.

The remaining part of the paper elaborates on these definitions. We identify hierarchical differentiation, anthropometric measurement and the principle of self-determination in science. We discuss subsequently logic, mathematics, physics (the metaphor of " motion "), biology (the metaphor of " body ") and economics (the metaphor of " value ").

SCIENCE IN THE ANTHROPOMETRIC STAGE

Logic and mathematics

Since hierarchical differentiation does not give rise to a hierarchical system which shows the property of the natural number system, logic and mathematics do not reach the level of " abstractness " we encounter today. They remain " captured " within the concrete everyday world. In Aristotelian syllogistic, formalization with the use of symbols that do not relate closely to concrete objects is not very advanced.

The relative " concrete " character of Aristotle's formalization follows in the first place from the " categorical restriction " imposed upon the " variables ". These variables cannot be " loosened " from the syllogistic figure, they perform the function of " term variables ". They are in fact " parking-places " for expressions to be filled in, in contrast to a totally formalized logical system like PL or PL1, where reference to an existing structure is totally redundant. The second restriction on Aristotelian syllogistic is even more striking : the terms to be filled in may not be empty. Non-existing entities therefore do not find a place in Aristotle's logical system, what accounts for the connection between syllogistic and concrete life[16].

Michel argues that Aristotelian syllogistic and Euclidean geometry contain a common inspiration[17]. Mathematics in general, and geometry in particular, are often encountered in the context of measurement — *geometrein* after all signifies " measurement of the earth ". Since anthropometric measures were not unified in an abstract system containing the attributes of number, it is no surprise that abstraction in mathematics remained within certain limits too.

The geometrical representation, which follows from the close connection between mathematics and human representation of the concrete earth, leads to the impossibility of certain magnitudes. The " discovery " of the irrational length $\sqrt{2}$ indeed signified a terrible shock for the religious inspired Pytha-

16. W. De Pater, R. Vergauwen, *Logica : Formeel en informeel*, Leuven, 1992, 97-99.
17. P.-H. Michel, *De Pythagore à Euclide*, Paris, 51.

gorean mathematicians[18]. Greek " space " should therefore be seen as representing the concrete human perception of the earth, which leads to an unchangeable and absolute character of " space ", in contrast to contemporary conventional definitions of space. Euclid's and Archimedes' works (at least those which have come down to us) do not even contain the word " space " : the Greek mathematician meant to discover a real order. Furthermore, mathematical demonstration relies heavily upon intuition, and at least partly on concrete drawings. Finally, Euclid's analysis starts with the concrete, and goes from there to abstract notions : he introduces first of all the circle, and only thereafter circumference. We conclude that the " concrete " attitude of the anthropometric stage does not give rise to a need of abstract, formalized systems. This attitude is reflected in anthropometric logic and mathematics, where the " variables " remain closely linked to the concrete world man is living in.

Physics : the Metaphor of Motion

The absence of abstract principles of measurement implies that the metaphors of measurement (motion, body and value) remain closely tied to each other. Mirowski discusses two examples from Aristotelian physics : the teleological explanation of movement and the specific character of the impetus theory[19]. Teleology connects the metaphors of motion and body, since movement is the process in which the specific essence of a body is realized. Natural movement is the tendency of a body to reach for its natural place. Aristotle discusses several teleological causes (organisms, artifacts, intentional actions or the nature of things), but also some teleological effects (processes, parts of bodies, things that emerge or are as they are)[20]. We therefore recognize two entirely different forms of theological causation which Aristotle does not differentiate from each other : intentional actions of agents and natural organisms (agency and function model). Aristotle writes : " Further, in any operation of human art, where there is an end to be achieved, the earlier and successive stages of the operation are performed for the purpose of realizing that end. Now, when a thing is produced by Nature, the earlier stages in every case lead up to the final development in the same way as in the operation of art, and vice versa, provided that no impediment balks the process "[21].

Aristotle states " as in agency, as in nature " and " as in nature, as in agency ", and he does not analyze the notion " for the purpose of ". According to Mirowski Aristotle did not differentiate intentional actions from natural processes ; only a contemporary reader regards the " missing " analysis of tele-

18. M.J. Greenberg, *Euclidean and Non-Euclidean Geometries*, San Francisco, 1974, 6.

19. P. Mirowski, *More Heat than Light, op. cit.*, 110-111.

20. D. Charles, " Teleological Causation in the Physics ", in L. Judson (ed.), *Aristotle's Physics*, Oxford, 1991, 101-28.

21. Aristotle, *Physics*, II, VIII.

ology as a failure in Aristotle's thought. The need to formulate an abstract theory in which movement and body or natural and intentional action are differentiated from each other, remains absent in Aristotle's context.

The same attitude is reflected in Aristotle's theory of the impetus. Since rest is the natural state of a body, it does not need further explanation ; and as movement keeps the body out of its natural position, a constantly operating impetus acts on the body. Since movement is defined as the realization of the specific essence of a body, the metaphors of movement and body are connected ; and since the " essences " of bodies cannot be represented in an abstract formal system, the same applies to the " movements " of these bodies, and therefore to the impetus acting on them. Vector addition and subtraction of forces is therefore absent in Aristotle's physics, because incommensurable things cannot be added. The principle of self-determination implies that the measures remain closely tied to the objects they are measuring.

Biology : the Metaphor of Body

The principle of self-determination is also present in Aristotle's classification of animals[22]. Aristotle's scientific attention is directed to concrete living organisms. This leads to an empirical notion of the concept of " species " in which hierarchical differentiation is reflected. Aristotle divides the animals in groups according to their specific features : presence or absence of certain body parts, or the form of these body parts. This methodology allows several different classifications, depending on the organ which is taken as a criterion. Using this methodology we establish a classification which reflects a general notion of " higher " and " lower " animals similar to the role of individuals in society, without a general comparison principle with the characteristics of the natural number system. A " higher " individual is not allowed to perform all actions which a " lower " individual is authorized to do ; and a " higher " animal does not possess all features of a " lower " animal.

The notion " species " does not play a theoretical part in Aristotle's biology : the (male) individual transmits his inheritable features to the next generation. The individual Socrates possesses several attributes of larger categories (man, animal), but these attributes are transmitted individually, without interference of an abstract notion " species ". This implies that an evolutionary theory does not exist within Aristotle's biological system, since this would require an abstract notion " species ". According to Aristotle a discussion of " species " would be redundant, because we would have to treat the same organs several times.

We argued above that the concepts of " motion " and " body " are not differentiated in the anthropometric stage. The motion of a body expresses its

22. P. Pellegrin, " Aristotle : a Zoology without Species ", in A. Gotthelf (ed.), *Aristotle on Nature and Living Things*, Pittsburgh, Pennsylvania, 1985, 95-115.

essence, the movement of body parts is directed to the essence of the body, and the body possesses those parts which are in accordance with its essence.

Economics : the Metaphor of Value

We will now argue that in economic " theory "[23] body and value are identified. Since " human inequality " forms the foundation of society, no abstract universal values exist, because this would imply the application of universal values in mutually incommensurable situations. The societal positions of the individuals involved in the exchange process have to be taken into account. Two goods, e.g. shoes and a house, are incommensurable, because they denote completely different objects. However, the final cause of both goods is the satisfaction of human needs ; this feature makes both goods commensurable. Since humans are unequal " by nature ", we cannot formalize these needs in an abstract framework : the " weight " of the " need " will depend on the societal position of the individuals involved in the exchange process[24]. Since the goods are incommensurable, they cannot be related to each other ; in the exchange process, the humans are related to each other through the medium of the good[25]. Money is only a convention, a medium for the exchange process[26]. Aristotle elaborates on the exchange process in his famous example of a master builder and a shoemaker : " As therefore a builder is to a shoemaker, so must such and such a number of shoes be to a house [or to a given quantity of food] ; for without this reciprocal proportion, there can be no exchange and no association "[27].

We observe that Aristotle relates builder and shoemaker, and not the *labour of* the builder and *the labour of* the shoemaker : this last relation can only be genuine in a context of human " equality ", in which the similarity of the different forms of human labour *as human labour* is recognized. The process of exchange is a movement, in which one individual offers a good to an other individual, which creates a disequilibrium : only the second person receives " utility ". To overcome this disequilibrium, the second individual has to offer another good in order to give " utility " to the first individual[28]. " Utility " here is a concrete concept which cannot be detached from the individuals involved,

23. Aristotle did not have an " economic theory " as we understand it today. His *Oikonomikos* is a collection of practical rules for the maintenance of land and farm.

24. The Austrian neoclassical school sometimes claims that Aristotle is their predecessor, since he also states that " utility " forms the foundation of value. However, because Aristotle's concept of " utility " is closely related to (the societal position) of the concrete individual, the contrast with the *abstract* utility theory of value of the Austrian school is striking. See L.J. Zimmerman, *Politieke economie van Plato tot Marx*, Groningen, Wolters-Noordhoff, 1987, 48.

25. J. Soudek, " Aristotle's Theory of Exchange ", *Proceedings of the American Philosophical Society*, 96 (1952), 45-75.

26. Aristotle, *Nichomachean Ethics*, 1133a, 25-31.

27. Aristotle, *Nichomachean Ethics, op. cit.*, 1133a, 20-25.

28. J. Soudek, " Aristotle's Theory of Exchange ", *op. cit.*

because it is related to the social status of both individuals. A " hierarchically higher " person will receive a larger share in the exchange process, because his " utility " is related to his " higher " essence. Body, motion and value are intertwined : the exchange process is a *movement* in which the proportion between the final causes of two objects (*value*) is equal to the proportion between the social status of the two individuals involved in the exchange process (*body*).

LA DIFFUSION DE LUCRÈCE EN ITALIE AU XVIe SIÈCLE[1]

Amalia PERFETTI

Dans une lettre retrouvée récemment par Tommaso Campanella à Peiresc, le philosophe de Stilo trace les lignes d'une tradition italienne de l'atomisme antique[2]. Bien que la pensée de Campanella, du point de vue de l'atomisme, ne présente pas un réel intérêt, cette lettre nous livre cependant un jugement très instructif sur la reprise de l'atomisme classique avant Gassendi. Par ailleurs, Campanella sachant que Gassendi travaillait sur un *Democrito* — la lettre est du mois de juin 1636 — écrivait à son correspondant français qu'il y avait aussi en Italie plusieurs " democritici ", comme il appelait les philosophes qui s'étaient occupés d'atomisme. Parmi ces derniers, se trouvait Galileo, à propos duquel il écrit : *in molte cose, massime nei principi, è con Democrito e dal discorrere c'ha fatto meco in Roma, e da quel che ne scrive nell'opuscolo " De natantibus " e nel " Saggiatore ", il padre Castelli et monsignor Ciampoli e condiscepoli così per tal lo difendono*, ou bien encore Paolo Sarpi et, pour l'époque de Telesio, Francesco Sopravia qui *leggeva a meraviglia a tutti l'opinione di Leucippo e Democrito*, et dont il avait lu les ouvrages manuscrits parmi lesquels se trouvait probablement un *De natura rerum*.

Si cette lettre de Campanella est sans doute un précieux témoignage replaçant les travaux de Gassendi à l'intérieur d'une très large réflexion sur l'atomisme antique entre la fin du XVIe siècle et les premières décennies du XVIIe siècle, la première traduction anglaise du *De rerum natura*, imprimée et partielle, nous offre, dans la même perspective, des aspects encore plus intéressants. L'auteur de cette traduction, publiée à Londres en 1656, est John Eve-

---

1. Cette communication présente l'approche générale et la ligne de recherche que je suis en train de développer dans ma thèse de doctorat en Histoire des sciences (Università degli Studi di Firenze).

2. G. Ernst, E. Canone, " Una lettera ritrovata : Campanella à Peiresc, 19 giugno 1636 ", *Rivista di storia della filosofia*, 49 (1994), 353-366. Des observations intéressantes sur cette lettre par rapport à l'atomisme de Galilée sont données par F. Favino, " A proposito dell'atomismo di Galileo : da una lettera di Tommaso Campanella ad uno scritto di Giovanni Ciampoli ", *Bruniana & Campanelliana*, 3 (1997), 265-282.

lyn[3] dont Kargon a bien souligné l'importance dans le développement de l'intérêt pour la philosophie atomistique en Angleterre à partir du milieu du XVIIᵉ siècle[4].

Connaisseur averti des ouvrages de Gassendi et de Charleton et lecteur attentif des commentaires du *De rerum natura*, J. Evelyn cite également dans son long commentaire au premier livre du poème de Lucrèce, des auteurs italiens pour lesquels, et surtout en ce qui concerne la philosophie de la nature, on peut penser à une influence directe et très forte de l'atomisme antique[5].

Mais, si les savants du XVIIᵉ siècle portèrent leur attention sur l'époque, qui, pour utiliser une expression de Robert Lenoble : "oubliera le monde de Dante et inclinera vers celui de Lucrèce"[6], on ne peut pas dire la même chose des historiens contemporains.

Trop souvent lus avec les yeux de chercheurs spécialistes des XVIIᵉ et XVIIIᵉ siècles, les travaux des premiers atomistes de l'époque moderne ou des premiers philosophes et naturalistes qui commencèrent à voir la philosophie de la nature de Lucrèce comme une alternative valable au cosmos d'Aristote, ont presque toujours été jugés trop éloignés des problématiques ouvertes par la révolution scientifique pour être effectivement intégrés pleinement dans cette renaissance de l'atomisme antique qui constitue sans doute un des aspects les plus intéressants du XVIIᵉ siècle.

Ma recherche est donc structurée comme une tentative pour circonscrire, principalement dans un premier temps du point de vue de l'Italie, cette problématique avant le *De vita et moribus Epicuri* et les *Animadversiones in Decimum librum Diogenis Laertii* de Gassendi[7]. Mes premiers résultats me conduisent déjà à affirmer que la lecture du *De rerum natura* de Lucrèce au XVIᵉ siècle n'était pas liée seulement à des problématiques morales ou philologiques comme la bibliographie critique le laisserait entendre. Et je pense ici, aussi bien à des ouvrages classiques sur l'atomisme[8], qu'à des travaux plus récents, comme celui d'Howard Jones sur la tradition épicurienne qui relègue la diffu-

3. J. Evelyn, *An Essay on the First Book of T. Lucretius Carus* " De rerum natura ", London, 1656.

4. R.H. Kargon, *Atomism in England from Hariot to Newton*, Oxford, 1966, 115-126. En particulier pour cette traduction voir, A. Perfetti, " J. Evelyn et " the rational Bruno " ", *Bruniana & Campanelliana*, 1 (1995), 233-248.

5. Parmi les auteurs italiens cités par J. Evelyn, il faut souligner la présence de M. Marullo, P. Candido, A. Paleario, S. Capece, F. Patrizi, G. Bruno, T. Tasso et T. Campanella.

6. R. Lenoble, *Histoire de l'idée de nature*, Paris, 1966, 277.

7. Les deux ouvrages de Gassendi furent publiés à Lyon — *apud G. Barbier* — respectivement en 1647 et en 1649. La bibliographie sur Gassendi est très vaste ; reste fondamentale la monographie de O.R. Bloch, *La philosophie de Gassendi. Nominalisme, matérialisme et métaphysique*, La Haye, 1971.

8. Voir, par exemple, K. Lasswitz, *Geschichte der Atomistik vom Mittelalter bis Newton*, Hamburg, Leipzig, 1890 (2 vols).

sion du philosophe de Samos au XVIe siècle en Italie dans un chapitre général dédié au débat ouvert par l'Humanisme[9].

En effet, dès la fin du XVe siècle — la découverte du manuscrit du *De rerum natura* date de 1417 mais sa circulation effective fut un peu plus tardive[10] — on peut déjà voir à côté de la première réception des vers lucrétiens dans leur contenu moral, un intérêt pour la philosophie de la nature épicureo-lucrétienne chez des auteurs qui, comme Ficino et Marullo, par exemple, avaient une vision de la *Natura*, pour reprendre les mots de Eugenio Garin : *animata, divinizzata, vestita di bellezza*[11]. Par ailleurs, c'est précisément dans le milieu florentin de Ficino, qui avait livré au feu son commentaire du *De rerum natura*[12], que nous trouvons probablement le premier ouvrage systématique consacré au poème lucrétien, en dehors des quelques notes manuscrites pour des éditions comme celles de Pontano[13]. Il s'agit d'une paraphrase des trois premiers livres du *De rerum natura* publié à Bologne en 1504 par le florentin Raffaele Franceschi[14]. Cet ouvrage, peu connu et sur lequel les historiens n'ont jamais vraiment porté leur attention est, à mon avis, caractéristique de la première diffusion du *De rerum natura*[15].

L'esprit de Franceschi par rapport au poème lucrétien est sans doute bien loin de celui des philologues et des poètes de son époque. Le but de sa paraphrase — et déjà le choix de cette forme littéraire est significative semble avoir pour objet de rendre plus accessible la lecture du *De rerum natura* à travers l'analyse des principes constitutifs du système lucrétien, en excluant tous les passages plus proprement poétiques, comme les vers consacrés à Venus ou les éloges adressés à Épicure.

En effet, sa *Paraphrasis* se concentre surtout sur les parties du poème liées à l'explication de la création, de la génération, du mouvement et de la concep-

9. H. Jones, *The epicurian tradition*, London, New York, Routledge, 1992 (in paperback, 1989 in hardback). Voir en particulier le chapitre 6, " The humanist debate ", 142-165.

10. *Cf.* R. Sabbadini, *Le scoperte dei codici latini e greci ne' secoli XIV e XV. Nuove ricerche*, Firenze, 1914, 80-82.

11. E. Garin, *Il ritorno dei filosofi antichi*, Ristampa accresciuta del saggio *Gli umanisti e le scienze*, Napoli, 1994, 100.

12. *Cf.* P.O. Kristeller, *Supplementum ficinianum*, I, Florentiae, Olschki, 1937, CLXIII (2 vols).

13. Les notes manuscrites de Pontano sont conservées dans les incunables du *De rerum natura*. *Cf.* C.A. Gordon, *A bibliography of Lucretius*, London, 1962, 51-52.

14. *Raphaelis Franci florentini in Lucretium Paraphrasis cum appendice de animi immortalitate*, Bononiae, Ioannem Platonidem Benedictorum, 1504.

15. La bibliographie sur Franceschi est en effet très limitée. La première nouvelle de cette paraphrase est parue dans C.A. Gordon, *A bibliography of Lucretius, op. cit.*, 53, 228-229. Sur Franceschi et sur sa paraphrase voir : S. Bertelli, " Notarelle machiavelliane. Ancora su Lucrezio e Machiavelli ", *Rivista storica italiana*, LXXVI (1964), 774-792, 782-784 ; U. Pizzani, " Dimensione cristiana dell'Umanesimo e messaggio lucreziano : *La Paraphrasis in Lucretium* di Raphael Francus ", *Validità perenne dell'Umanesimo*, a cura di G. Tarugi, Firenze, 1986, 313-333 ; reste toutefois fondamentale, même pour le milieu de Franceschi, l'oeuvre monumentale de A.F. Verde, *Lo Studio Fiorentino, 1473-1503*, V, Firenze, 1973-1994, III, II (1977), 848-849 et IV, III (1985), 1458-1460.

tion des atomes, conception à laquelle sont dédiées différentes pages où elle se trouve expliquée longuement et d'une façon claire. Parmi les problématiques lucrétiennes que Franceschi aborde je veux au moins rappeler rapidement sa *digressio* sur l'infinité dc l'univers. Cette thématique qui, comme Michel Blay l'a souligné dans sa conférence plénière, est exemplaire pour saisir les développements de la science et de la philosophie de l'époque classique[16], nous montre parfaitement l'attitude de Franceschi dans son travail sur Lucrèce. Comme cela était typique à l'époque, l'humaniste florentin met en relation les passages de Lucrèce avec les ouvrages d'Aristote et de Platon ; cependant l'ambiguïté avec laquelle, comme dans le cas de l'infinité de l'univers, il mélange l'*auctoritas* de ces sources avec les doctrines du poète latin nous laisse croire à une tentative pour montrer l'efficacité ou la séduction qu'exercent certaines thématiques lucrètiennes. De même, les références qu'il fait à son probable maître Ficino — surtout à la *Théologie platonicienne* et au commentaire du *Timée* — pour résoudre les questionnements ouverts par la lecture du *De rerum natura*, apparaissent comme des références à double sens puisque nous trouvons dans sa *Paraphrasis* longuement développée les doctrines épicureo-lucrétiennes rapidement rappelées par Ficino. Cela confirme notre impression suivant laquelle Franceschi voulait faciliter la compréhension des parties du poème de Lucrèce qui commençaient par être connues le plus souvent de façon indirecte.

Si la seule lecture de la *Paraphrasis* ne nous permet pas de penser que Franceschi puisse avoir été un lucrétien, le jugement de ses contemporains est sûrement précieux. Surnommé " Celatone ", celui qui se cache ou celui qui utilise la *celata* — sorte de casque — pour se protéger, il fut appelé dans une lettre de 1524 qui annonçait la nouvelle de sa mort — à l'époque où il enseignait *physica* au *Studio* de Pise — philosophe lucrétien : *Pluto* — nous pouvons lire dans cette lettre — *ille princeps inferum, de eius adventu multo gaudio fuit. Epicurus, Lucretius et plurimi philosophi obviaverunt ei ...*[17].

Les années qui suivirent la publication de la paraphrase de Franceschi virent un intérêt croissant pour le texte du *De rerum natura.* Ce sont en 1511 et 1512 les deux premières éditions philologiquement importantes du poème : la première publiée par Giovan Battista Pio chez le même éditeur bolognais que celui de Franceschi, Platone de Benedetti ; et la deuxième par Pietro Candido, publiée à Florence chez Giunta et dédiée, comme la *Paraphrasis*, au florentin Tommaso Soderini. Ces éditions, comme celles qui se multiplièrent aux XVIe et XVIIe siècles, méritent d'être étudiées non seulement d'un point de vue philologique, ce qui n'est pas mon but, mais plutôt pour l'écho que l'on peut en

16. M. Blay, " La science classique 'revisitée' ", *History of Modern Physics* : proceedings of the 20th International Congress of History of Science, Liège 20-26 July 1997, vol. XIV, in H. Kragh, G. Vanpaemel, P. Marage (eds), Turnhout, Brepols, 2002, 17-32 (De Diversis Artibus, Tome 57, N.S. 20).

17. A.F. Verde, *Lo Studio Fiorentino, 1473-1503,* II, Firenze, 1973, 377.

trouver dans les notes des discussions physiques et philosophiques de l'époque.

Parmi les autres textes, outre les éditions témoignant de la diffusion de Lucrèce dans la première moitié du XVIe siècle, on trouve par exemple le *Syphilis sive morbus gallicus*[18] de Girolamo Fracastoro et le *Zodiacus vitae* de Palingenio Stellato[19].

Le Morbus gallicus, pour lequel Pietro Bembo écrivit au même Fracastoro : (...) *parmi che scriviate in verso cose, tolte di mezzo la filosofia, molto poeticamente e molto più graziosamente che non fa Lucrezio molte delle sue*[20], a par rapport à la diffusion du *De rerum natura* un double intérêt. Non seulement Fracastoro choisit d'écrire en vers comme Lucrèce, mais il reprend aussi sa théorie des *semina* pour expliquer la contagion ; cette thèorie souligne un aspect très important de la diffusion de l'atomisme chez les médecins de la Renaissance[21].

Le rapport de Palingenio avec ses sources anciennes est plus compliqué. En effet, dans son ouvrage l'influence des idées platoniciennes et néo-platonicienne est très importante. En outre, s'il est bien clair que le *De rerum natura* constitue aussi pour lui un modèle littéraire très important, les pages du *De immenso* de Giordano Bruno nous en montrent ses limites par rapport à la structure de l'infini qu'il ne concevait pas avec la réalité physique qui est celle de l'atomisme de Lucrèce tel qu'il est repris par Bruno à la fin du siècle[22].

Si les ouvrages analysés jusqu'ici nous montrent déjà des aspects significatifs de la diffusion du *De rerum natura*, au milieu du XVIe siècle il y a un autre ouvrage particulièrement représentatif du rôle joué dans la réflexion sur le *principia prima* par les théories épicureo-lucrétiennes : il s'agit du *De principiis rerum* du napolitain Scipione Capece[23], et pour lequel, comme dans le cas de Fracastoro, Bembo parlera d'un style tout lucrétien[24]. Bien au-delà des jugements de Bembo c'est le même Capece, qui déclare son admiration pour le poète latin, le *pater Lucretius* qui, chez les *italici* avec un *blandus cantus*

18. G. Fracastoro, *Syphilis sive morbus gallicus*, Vérone, 1530.

19. De P. Stellato, *Zodiacus vitae, ca.* 1534 et le 1536 est sortie une édition avec traduction française : Palingène, *Le zodiaque de la vie (Zodiacus vitae), XII livres*, texte latin établi, traduit et annoté par J. Chomarat, Genève, 1996.

20. La lettre est du 26 novembre 1525 et on peut la lire maintenant dans P. Bembo, *Lettere*, IV, edizione critica a cura di E. Travi, Bologna, 1987-1993, 315-317, 316.

21. *Cf.* J. Roger, *Les sciences de la vie dans la pensée française au XVIIIe siècle*, Paris, 1993, 121-140.

22. Pour la critique de Bruno à Palingenio voir le livre VIII du *De immenso et innumerabilis*, G. Bruno, *Opera latine conscripta*, publicis sumptibus edita, recensebat F. Fiorentino [F. Tocco, H. Vitelli, V. Imbriani, C.M. Tallarigo), I-III, Neapoli-Florentiae, Morano-Le Monnier, 1879-1891, I, II (1884), 286-318 (8 vols).

23. Scipione Capece, *De principiis rerum*, Venise, P. Manuzio, 1546.

24. Ce jugement se trouve dans la lettre de Bembo publiée dans la première édition du *De principiis rerum*.

expliqua la philosophie des grecs sur le *semina mundi*, et nous montre son intérêt pour le *De rerum natura*. Si tout d'abord, la forme littéraire de son ouvrage s'inscrit dans une tradition de poésie scientifique qui considérait le *De rerum natura* comme modèle, dont il pouvait trouver en outre un exemple significatif dans l'*Urania* de son maître Pontano[25], il y a ensuite d'autres éléments qui placent le *De principiis rerum* de Capece comme un des premiers ouvrages dans lesquels on trouve des points de contacts entre les idées épicureo-lucrètiennes et l'effort pour élaborer une nouvelle image du monde.

Sa réflexion sur les principes des choses se développe à partir d'un très long exposé de la philosophie de la nature atomistique qui occupe les 350 premiers vers de son poème et qui pose les caractéristiques fondamentales des *principia rerum* avec l'aide desquelles il critiquera les autres théories des éléments. En effet, s'il propose comme premier principe de toutes les choses l'*aër* et non pas les atomes c'est en s'appuyant sur de nombreuses définitions données par Lucrèce des *semina rerum* que le naturaliste napolitain explique sa conception de l'univers. Sans vouloir faire, comme son commentateur du XVIIIe siècle, de Capece un précurseur, il faut néanmoins remarquer que pour l'époque de sa publication on trouve dans le *De principiis rerum* des théories cosmologiques d'un certain intérêt. Avant tout le fait que Capece utilise de nombreux vers de son poème pour critiquer la distinction aristotélicienne entre l'éther et la matière du monde sublunaire. Les conséquences de l'homogénéité de la matière qu'il propose avec clarté et détermination sont évidentes : corruptibilité et fluidité des ciels liées à une nouvelle conception du mouvement des astres ; thèmes entre autres, qui dans les derniers temps ont été l'objet d'importantes études par Grant et Lerner[26]. Capece nie, en effet, la possibilité d'un mouvement circulaire parfait, en parlant de différents types de mouvement strictement liés au poids des corps, tant pour le ciel que pour le monde sublunaire pour lesquels il n'hésite pas à reprendre des thématiques du *De rerum natura*. Un exemple intéressant dans ce sens est fourni par la reprise des vers lucrètiens sur le mouvement des astres — V[e] livre — le conduisant à une sorte de théorie des tourbillons avec laquelle il justifie la *propria natura* du mouvement des *primordia rerum* et des astres.

Toutes ces analyses de Capece visent à bien expliquer son image du monde dans laquelle, il n'y a plus un monde céleste royaume de l'incorruptibilité, d'une harmonie des orbites circulaires et régulières et enfermé dans des limites précises. En effet Capece ne parviendra pas à parler de l'infinité de l'univers comme Lucrèce et plus tard Bruno, aussi s'il pense que l'on ne peut pas donner au monde une *certa mensura*, idée pour laquelle il sera par ailleurs cité par J.

---

25. Ce poème fut publié par A. Manuzio à Venise en 1505.

26. E. Grant, *Planets, Stars, and Orbs. The Medieval Cosmos. 1200-1687*, Cambridge, 1994 ; M.P. Lerner, *Le monde des sphères*, II, Paris, 1996-1997. Ces deux ouvrages, fondamentaux dans l'histoire de la cosmologie, ne dédient pas, de toute façon, beaucoup de place à la théorie atomistique de la matière.

Evelyn dans son commentaire du *De rerum natura* dont nous avons précédemment parlé[27].

Pour en terminer avec Capece je crois que son ouvrage, presque totalement ignoré par les historiens[28], peut trouver une place très importante dans la diffusion des idées épicureo-lucrètiennes, tant pour le fait qu'il est lié à son exigence d'aider à la compréhension de la nature du ciel, que par celui qu'il chante dans son poème la nécessité qu'il y a de faire des découvertes nouvelles. En outre, il faut rappeler que la diffusion du *De principiis rerum* ne fut pas du tout négligeable : six éditions au XVIe siècle, une au XVIIe, publiée en 1631 avec le *De rerum natura*, et deux au XVIIIe siècle, la première en 1754, fut publiée par le même abbé bénédictin, Francesco Maria Ricci, qui traduira aussi l'*Anti-Lucrèce* de Polignac. Ces dernières éditions témoignent d'une diffusion de l'ouvrage de Capece liée sans doute aussi à celle du poème de Lucrèce[29].

Dans la deuxième partie du XVIe siècle, mais je ne dirais sur cette période que quelques mots, l'intérêt pour le *De rerum natura* se développe : en porte témoignage de nouvelles éditions avec de longs commentaires, notamment celles de Denis Lambin[30] et de Hubert van Giffen[31], ou encore, d'un point de vue plus littéraire, la réflexion très articulée sur la poétique du *De rerum natura*, ainsi que sur la validité de la poésie scientifique, dont comme nous l'avons déjà vu, le poème de Lucrèce représente un modèle idéal.

Bien sûr, de nombreux autres auteurs interviendront dans ma recherche, ainsi et principalement Giordano Bruno, dont, à mon avis, on n'a pas encore parfaitement souligné l'importance de l'atomisme dans l'élaboration de sa pensée[32].

Même si les perspectives que j'ai présentées ici sont essentiellement celles d'un *work in progress*, j'espère néanmoins être arrivée à éclaircir quelques aspects de la diffusion de l'atomisme au XVIe. J'ai préféré faire un choix chronologiquement limité, mais qui m'a permis de dégager des développements possibles. En effet, même s'il faut rappeler que chez les auteurs dont j'ai parlé aujourd'hui, les théories de Lucrèce sont souvent mêlées avec d'autres sources anciennes, on y percevait déjà un intérêt pour l'atomisme fortement lié à la philosophie de la nature et à la nouvelle image du monde qui se développera

---

27. J. Evelyn, *An Essay on the First Book of T. Lucretius Carus* " De rerum natura ", *op. cit.*, 164.

28. Le seul article en partie dédié à la cosmologie de Capece est celui de F. Bacchelli, " Sulla cosmologia di Basilio Sabazio e Scipione Capece ", *Rinascimento*, XXX (1990), 275-292.

29. Pour la liste des éditions du *De principiis rerum*, voir C.A. Gordon, *A bibliography of Lucretius, op. cit.*, 296.

30. Lucretius Carus, *De rerum natura*, Paris, Denis Lambin, 1563.

31. *Cf.* Lucretius Carus, *De rerum natura*, Anvers, Hubert van Giffen, 1565 ; C.A. Gordon, *A bibliography of Lucretius, op. cit.*, 78-87.

32. J'ai présenté les premiers résultats de cette recherche sur Bruno dans : A. Perfetti, " Motivi lucreziani in Bruno : la Terra come *madre delle cose* e la teoria dei *semina* ", *Letture Bruniane del Lessico Intellettuale Europeo (1996-1997)*, Pisa, Roma.

pendant la Renaissance, pour s'épanouir, et pour ne rappeler que deux exemples importants, dans les pages des ouvrages de Gassendi ou dans celles du
galiléen Alessandro Marchetti qui traduira, en *versi toscani*, le *De rerum
natura*[33].

---

33. La traduction de Marchetti fut publiée à Londres, " Per Giovanni Pickard ", en 1717, elle
avait connu toutefois à partir de 1670 une diffusion manuscrite très importante. Sur Marchetti et
sur son *Lucrèce*, voir : M. Saccenti, *Lucrezio in Toscana*, Firenze, 1965.

Bacon and Eclecticism

Stephen Gaukroger

Bacon and Eclecticism

Eclecticism was a very potent force in seventeenth century natural philoso phy. We find it in major figures such as Boyle and Newton, and in minor ones, such as Digby and Charleton. We find it in unorthodox figures like Dee and Fludd, and in orthodox ones like Cudworth and More. We find defences of English natural philosophy in continental Europe explicitly framed in terms of its eclecticism, as in Jacob Sturm's *Philosophia Eclectiva* (1686), and later in the article on eclecticism in Diderot's *Encyclopédie*.

My concern today is with Francis Bacon. In Bacon's writings, there is often a fluidity, an eclecticism, and an unwillingness to decide between what seem to be competing views, trying instead to balance them, as if in deciding between them we would inevitably lose something valuable in the view we decided against. Although eclecticism is something that disappeared from seri- ous philosophy in the 18th century at the earliest, it is something so alien to modern philosophy that it has effectively been dismissed as a mode of pursu- ing philosophy. Moreover, it runs completely counter to the interpretative can- ons of modern historiography : it is bad enough to take things out of context, but then to attempt to combine them with things removed from totally different contexts looks like little more than pastiche. The trouble is that one thing the interpretative canons of modern historiography are supposed to secure is a bet- ter contextual understanding, and we will not achieve this simply by condemn- ing the kind of eclecticism that we find in 17th-century natural philosophy. Somehow, we need to come to terms with it in its own right. It is not just that no real understanding of Bacon is possible without it, but that no real under- standing of the bulk of English natural philosophy in the 17th century is possi- ble without it. Bacon is working within a current of thought that not only has a Renaissance tradition behind it, but which extends further into the early mod- ern era than many commentators would like to admit.

In this paper I want to look at three sources of eclecticism in Bacon's natural philosophy, while ranging freely over the 16th and 17th century context in which these sources arise and flourish. These are :

1. the *via media* approach to natural philosophy ;
2. the commonplace book tradition ; and
3. the " unity of truth " doctrine.

A *VIA MEDIA*

A crucial ingredient in the reform of natural philosophy for Bacon is a reform of its practitioners : if we neglect this element in his programme we will fail to see what was its practical cutting edge[1]. In this respect, Bacon's concerns can be seen as part of a general concern with the reform of behaviour which began outside scientific culture but which was rapidly internalized in English natural philosophy in the 17th century. A particular way of pursuing natural philosophy was associated with what can only be called a particular form of civility. The investigation of natural processes — observation and experimentation — was contrasted with and pitted against verbal dispute, the first being construed as a procedure by which we actually learn something, the second as consisting of mere unproductive argumentation for its own sake. In the context of English thought in the early modern era, the advocacy of experiment over scholastic disputation, and the advocacy of a " civil " approach in which some form of compromise is sought in scientific and philosophical matters, are indissolubly linked[2]. One crucial thing at stake in both is a rejection of scholastic disputation : it is both the wrong way for science to be pursued and the wrong way for scientists to behave. The key idea is that civility and good sense dictate that one should pursue a *via media*, some form of middle position which both parties to a dispute could accept.

Boyle is a good example of this linking of the appropriate form of scientific practice with the behaviour appropriate to the scientist. There is a constant attempt in Boyle to find a *via media* in metaphysical disputes. The corpuscular hypothesis, he tells us, is something that transcends metaphysical disputes between the Cartesian and Epicurean schools, whose hypotheses " might by a person of a reconciling disposition be looked on as... one philosophy "[3]. Eclecticism is presented here as an ingredient in gentlemanly behaviour, some-

---

1. Two recent accounts of Bacon's reforms have drawn attention to this aspect of his programme : J. Martin, *Francis Bacon, the State, and the Reform of Natural Philosophy*, Cambridge, 1992 ; J.E. Leary Jr., *Francis Bacon and the Politics of Science*, Ames, Iowa, 1994.

2. See the discussion of the " gentlemanly " mode of argument in S. Shapin, *A Social History of Truth : Civility and Science in 17th Century England*, Chicago, 1994 ; S. Shapin, S. Schaffer, *Leviathan and the Air Pump : Hobbes, Boyle, and the Experimental Life*, Princeton, 1985.

3. Preface to *Some Specimens of An Attempt to make Chymical Experiments useful to illustrate the notions of the Corpuscular Philosophy*, 1, Boyle, 1772, 355-358 ; quotation from p. 356. I am indebted to P. Anstey for references to Boyle's works.

thing to be contrasted with the adversarial mode of scholastic disputation. Boyle is possibly developing a theme in Bacon here, for Bacon himself explicitly defends the *via media*, using the images of steering between Scylla and Charybdis, and the flight of Icarus, in his *De Sapientia Veterum*, telling us that " Moderation or the Middle Way is in Morals much commended, in Intellectuals less spoken of, though not less useful and good "[4]. And Bacon's theory of " method ", as well as being designed to increase human collective power to discover natural laws and manipulate natural processes, was also intended, as a means to achieving this power, to provide a strict regimen which continually curbed the spontaneous tendencies of the mind. It is a theory about how to shape scientists, so that (contrary to their natural inclinations) they manifest the requisite good sense and behaviour in their observation and experiment. Avoiding extremes is important here — to avoid the " Idols of the Cave ", for example, we must steer a middle course between " extreme admirations for antiquity " and " extreme love and appetite for novelty "[5] — and it is indicative of the fact that Bacon's proposals are as much about reforming behaviour as about following productive procedures.

COMMONPLACE BOOKS

We can find mild and strong forms of eclecticism in the 16th and 17th centuries, and both play some role in Bacon's approach. The " mild " form of eclecticism I have in mind has its roots in a genre particularly common in the Renaissance, that of the commonplace book. The best known use of such commonplace books is in the realm of rhetoric, and they were places in which one jotted down passages, arguments, factual information, or turns of phrase that particularly struck one, although there are medical, travel, recipe, and other forms of such commonplace books common through the Middle Ages and Renaissance.

Natural philosophy was often pursued through such books in the Renaissance, and the model was Pliny's *Natural History*, which mixes myth, observation, and hearsay as well as information condensed from (by Pliny's own estimate) 2.000 books[6]. Pliny's *Natural History* is one of the few ancient sources that Bacon praises and although the genre of natural philosophy commonplace books has been relatively neglected until recently, it appears to have been a popular one, with Bodin's *Universae naturae theatrum*[7] providing a

4. *Works*, VI, 754.

5. *Novum Organon*, I, lvi ; *Works*, I, 170/iii, 59-60.

6. See the discussion of Pliny in G.E.R. Lloyd, *Science, Folklore and Ideology : Studies in the Life Sciences in Ancient Greece*, Cambridge, 1983, 135-149 ; R. French, *Ancient Natural History : Histories of Nature*, London, 1994, ch. 5.

7. Bodin, *Universae naturae theatrum*, 1596.

Renaissance version of Pliny[8]. As one commentator has pointed out, Bodin used the commonplace book as an arsenal of " tidbits of knowledge which he divorces from their original context in order to suit his own purposes "[9]. Although the use for which he intends them is different, Bacon too occasionally pursues natural philosophy by means of such commonplace books, most notoriously in his *Sylva sylvarum*, which throws together materials from the same kinds of disparate sources that we find in Pliny, without regard for consistency[10]. Nor is this an unconscious usage on Bacon's part, for in the *Advancement of Learning*, he offers an explicit defence of commonplace books, provided the collection of materials from them can be reformed along his own lines : " there can hardly be anything more useful even for the old and popular sciences, than a sound help for the memory ; that is a good and learned Digest of Common Places. (...) I hold diligence and labour in the entry of common places to be a matter of great use and support in studying ; as that which supplies matter to invention, and contracts the sight of the judgement to a point. But yet it is true that of the methods and frameworks of common places which I have hitherto seen, there is none of any worth ; all of them carrying in their titles merely the face of a school and not of a world ; and using vulgar and pedantical divisions, not such as pierce to the pith and heart of things "[11].

Nevertheless, in many respects, the role of the commonplace book is the most difficult of the models for Bacon's natural philosophy to assess. In particular, we must be careful not to run together various uses of commonplace books, and Bacon's indebtedness to the genre does not extend as far as the attempt to build up one's arguments from a patchwork of quotes and borrowings from ancient sources. This subgenre, known in the early 17th century as *cento*, finds notable exponents in Lipsius — who tells us that he aims to offer

8. See A. Blair, " Humanist Methods in Natural Philosophy : The Commonplace Book ", *Journal of the History of Ideas*, 53 (1992), 541-551 ; The persistence of the commonplace book and the seriousness with which it continued to be taken in the seventeenth century is indicated by the fact that J. Locke's first published original work was *A New Method of Making a Common-place Book*, which introduced a new means of indexing the contents of such books (by first consonant and first vowel). The first appearance of the piece was in French translation in J. Le Clerc's *Bibliothèque universelle et historique* in 1686. On the later history of commonplace books and their connection with natural science see R. Yeo, " Ephraim Chamber's *Cyclopedia* (1728) and the Tradition of Commonplaces ", *Journal of the History of Ideas*, 57 (1996), 157-175.

9. A. Blair, " Humanist Methods in Natural Philosophy : The Commonplace Book ", *op. cit.*, 545.

10. See the discussion in G. Rees, " An Unpublished Manuscript by Francis Bacon : *Sylva sylvarum* Drafts and Other Working Notes ", *Annals of Science*, 38 (1981), 377-412.

11. *Adv Learn*, Part 5, ch. 5. *Works*, III, 435. The use of commonplaces is not discussed in any detail in Bacon, although a fragment " Of the Colours of Good and Evil ", which is appended to the 1597 edition of the *Essays* (*Works*, VII, 77-92), and reappears in 1623 as part of Book 6 of *De Augmentis* (*Works*, I, 674-88), does deal with the question briefly, and his term " colours " refers to general precepts of argument, or commonplaces ; See L. Jardine, *Francis Bacon : Discovery and the Art of Discourse*, Cambridge, 1974, 219. *Cf.* Q. Skinner, *Reason and Rhetoric in the Philosophy of Hobbes*, Cambridge, 1996, 115-116.

instruction " not by my owne sayings, but by the precepts of ancient authors, delivered also in their own wordes "[12] — and in Burton, who starts out by telling us that : " As a good house-wife out of diverse fleeces weaves one piece of cloth, a bee gathers wax and honey out of many Flowers, and makes a new bundle of all, (…) I have laboriously collected this cento out of various writers (…) [but] I cite and quote mine authors (…) *sumpsi, non surripui* "[13].

This genre is not wholly alien to Bacon, as those accounts which stress Bacon's opposition to earlier thought might have us believe, but nor is it his *modus operandi*. Sometimes commonplaces seem to function as the skeleton on which he builds even if this building is minimal, as in the short aphoristic writings and in *Sylva sylvarum*. On other occasions, as in the *Novum Organon* or the core material for the *Great Instauration*, there seems to be quite a schematic skeleton which is fleshed out in terms of commonplaces (in the form of uncontentious information selected from ancient authors, for example), amongst other things.

THE UNITY OF TRUTH

A very different and far stronger form of eclecticism is what might be called the " unity of truth " or " unity of knowledge ", which is manifested in a number of different ways. Part of what is involved is the idea that true knowledge would bring together all the different pursuits of knowledge, from natural philosophy to ethics, from grammar to theology. Such a conception was commonplace in the 16th and 17th centuries, and it was almost always pursued with some degree of eclecticism. The eclecticism becomes inescapable when one maintains that some of these pursuits are incommensurable with one another. This happened most clearly in disputes over the claims of reason and revelation, or between philosophy and theology, disputes that flared up throughout the 16th and 17th centuries. One of the most problematic of these was the doctrine of " mortalism " advocated by the Paduan philosopher, Pietro Pomponazzi, and condemned by the Lateran Council in 1513. Pomponazzi maintained that, philosophically speaking, the soul was the form of the body, and there was no such thing as an uninstantiated form, so that the death and corruption of the body resulted in the disappearance of the soul. On the other hand, he accepted on faith the Church teaching of the personal immortality of the soul. We are inclined to say that both conclusions cannot be right, but Pomponazzi treated the problem more along the lines of a dilemma : two completely different lines of thought, both of which he treated as completely compelling and

12. Justus Lipsius, *Sixe Books of Politickes or Civil Doctrine*, in W. Jones (transl.), London, 1594, Cited in Q. Skinner, *Reason and Rhetoric in the Philosophy of Hobbes*, op. cit., 118, who discusses the genre.

13. R. Burton, *The Anatomy of Melancholy*, 1, London, 1826, 11 (2 vols).

neither of which he was prepared to renounce, led to incompatible conclusions. Somehow one must embrace both[14].

This is an extreme case, where doctrines which will never be reconciled are accepted as both expressing some aspect of a single truth, but there are many cases in the grey area between these and those cases where one expects or hopes for eventual reconciliation. Such an intermediate case is Newton, for whom systems that might otherwise be seen as alternatives are seen as complementary, as each making a legitimate claim to provide access to some part of the truth, for truth itself lay beyond the apparently conflicting representations of it that were currently available. As Dobbs points out, in the course of his life Newton, " marshalled the evidence from every source of knowledge available to him : mathematics, experiment, observation, reason, the divine revelations in biblical texts, historical records, mythology, contemporary scientific texts, the tattered remnants of ancient philosophical wisdom, and the literature and practice of alchemy "[15]. In the case of Newton, this probably has as one of its sources the balance of approaches adopted in the interpretation of biblical prophecy, where, in the work of Henry More for example, the fact that any method of interpretation one might use was susceptible to error led to the advocacy of several approaches, not so much despite their incompatibility but because of their incompatibility : in that way one can be more confident that one has missed nothing[16]. What we end up with is a form of eclecticism in which doctrines are brought together which not only have no relevant points of contact for us, but where it is often difficult, if not impossible, to reconstruct what the points of contact might have been for the author who brought them together. One finds writers selecting freely from among available doctrines those bits that seem to be weldable together into some new synthesis/combination which somehow either brings one closer to the truth, or at least does not close off any possible avenues. It is as if one must fill in the gaps with something, even if there is no obvious connection between them : indeed, there being no obvious connection between them was often an incentive to assume that such a connection must lie at an especially deep level.

14. See the discussion in P.O. Kristeller, *Renaissance Thought and its Sources*, New York, 1979, ch. 11. On the details of the Pomponazzi case see G. Saitta, *Il Pensiero Italiano Nell' Umanesimo e Nel Rinascimento*, II, Bologna, 1949-1951, ch. 4-6 (3 vols) ; E. Gilson, " Autour de Pomponazzi : problématique de l'immortalité de l'âme en Italie au début du XVIe siècle ", *Archives d'histoire doctrinale et littéraire du Moyen Âge*, 18 (1961), 163-279.

15. B.J. Teeter Dobbs, M.C. Jacob, *Newton and the Culture of Newtonianism*, New Jersey, 1995, 9. Dobbs offers more detailed discussion in her *The Janus Face of Genius : The Role of Alchemy in Newton's Thought*, Cambridge, 1991, esp. ch. 1, and in the " condensed " version of this chapter which appeared as " *The Unity of Truth* : An Integrated View of Newton's Work ", in P. Theerman, A.F. Seeff (eds), *Action and Reaction : Proceedings of a Symposium to Commemorate the Tercentenary of Newton's Principia*, Newark, 1993, 105-122.

16. See R.H. Popkin, " The Third Force in XVIIe-Century Philosophy : Scepticism, Science, and Biblical Prophecy ", *Nouvelles de la République des Lettres*, 1983, 35-63.

The " mild " form of eclecticism is effectively a form of juxtaposition whereby one attempts to discern a pattern by placing everything side by side. The " strong " form, on the other hand, is quite different, and rests on the idea that questions can be contextualised in different ways, and that different contextualisations may yield incommensurable approaches and solutions. The extent of eclecticism in Bacon is beyond doubt : as Brian Vickers has put it, Bacon was " an incorrigible addict of modes of thinking which his expressed programme would replace : allegory, myth, iconographical symbolism, alchemy "[17]. The problem is what kind of eclecticism this is. It would be very convenient to assume that Bacon's eclecticism arises from a juxtaposition of unrelated elements, so that the aim of the exercise would then be to discover whether there were, perhaps, deep relations between them which we had missed. But two factors indicate that we should be cautious. First, Bacon occasionally makes remarks — such as his claim that " the more discordant therefore and incredible any Divine mystery is, the more honour is shown to God in believing it "[18] — that suggest he believes that there are some ultimate truths that may lie beyond our comprehension, and that lack of fit with experience is not an appropriate criterion by which to judge such truths. Second, the question of the kind of eclecticism at issue is not a marginal one, and it is very important that we get it right, because contextualisation plays a crucial role in Bacon. The historical contextualization of philosophical positions, the attempt to show that purported statements of universal truths are in fact something constrained by, and a product of, a particular historical context, is, after the very earliest writings a major form of criticism of earlier authors in Bacon, and one which was to be partly constitutive of modern thought in the Enlightenment — e.g. in the attempts of Voltaire and his successors to relativize Christianity to a particular (primitive) culture — and the Romantic era, where it was freed from its negative, purely critical role and became a way of understanding the development of cultural forms such as philosophy itself.

There is a sense in which what is at stake in Bacon's project is understanding rather than knowledge. This is not quite the distinction between practical and theoretical knowledge, and it is certainly not the distinction between technology and science. There is, for us, no easy route to the appreciation of what Bacon was attempting to provide. What we need to understand is how and why he attempts to transform the largely esoteric discipline of natural philosophy into a public practice, why he tries to achieve this largely through the reform of the mentality of the practitioners of these disciplines, and how his legal model for reform works both to enable him to think through, and facilitate the execution of, this process. In pursuing these questions, I believe we can gain

17. B Vickers (ed.), *Essential Articles for the Study of Francis Bacon*, Hamden, Conn., 1968, xviii.
18. Cited in H. Haydn, *The Counter-Renaissance*, New York, 1960, 10.

some insights into the search for a form of understanding which fired the imagination of Bacon's successors for two centuries and more, but which has been lost sight of in the modern notion that, ultimately, the only rationale for science is its discovery of the truth[19].

19. On the role of truth, see my " Justification, Truth, and the Development of Science ", *Studies in History and Philosophy of Science*, 29 (1998), forthcoming.

MECHANICISM, ORGANICISM, HOLISM IN LE BON'S CONCEPT OF THE CROWD : A HERMENEUTICAL INTERPRETATION OF PARADIGM COMBINATION

Oleg I. GUBIN

> " Who is it that in their intercourse with especially gifted people does not make the effort to listen between the words, just as read between the lines of rich and complex texts... " (Friedrich Schleiermacher, *Herme-neutik und Kritik*)

More than one hundred years have passed since Dr Le Bon published *Psychologie des foules* (*The Crowd*[1]) which became " one of the best-selling scientific books of all times "[2]. What intrigues in *Psychologie des foules* is the " transformation explanation "[3] of crowd's " demeaning influence "[4] on individuals. As contrasted to rational individuals, members of the crowd display " collective regression "[5] and " primitive impulses "[6], " deindividuation "[7] and " collective uniformity "[8], and a loss of " rational faculty and of moral sense "[9]. Scholars suggested different reasons of Le Bon's " transformation explanation ". His opponents, for example, sociologists Gamson and Ober-

1. G. Le Bon, *The Crowd*, 2nd printing, New Brunswick, [1895] 1997, 232.

2. R.A. Nye, " At the Crossroads of Power and Knowledge : The Crowd and Its Students ", *Introduction to the Transaction Edition in G. Le Bon, The Crowd*, New Brunswick, 1997, 11.

3. C. McPhail, *The Myth of the Madding Crowd*, New York,1991, 19.

4. Ch. Lindholm, *Charisma*, Cambridge, 1990, 40.

5. E.G. King, *Crowd Theory as a Psychology of the Leader and the Led*, Lewiston, 1990, 44.

6. J.A. Schumpeter, *Capitalism, Socialism and Democracy*, London, [1943] 1976, 257.

7. C. Kelly, S. Breinlinger, *The Social Psychology of Collective Action : Identity, Injustice and Gender*, London, 1996, 36.

8. E.G. King, *Crowd Theory as a Psychology of the Leader and the Led, op. cit.*, 43.

9. A. Oberschall, *Social Movements : Ideologies, Interests, and Identities*, New Brunswick, 1993, 5.

schall, see the reason in Le Bon's antidemocratism, misinterpretation of the
18[10]-19[10] centuries collective actions, and wrong methodology[10]. Contrary to
them, historians of science Nye, van Ginneken, Barrows and Moscovici praise
Le Bon for elaborate generalization on mass behavior during the French Revolution, Paris Commune, and Dreyfus affair[11]. Social psychologists Lindholm,
King and McPhail discovered in Le Bon's research the influence of suggestion-
imitation and other socio-psychological theories[12]. This paper offers rather different, *scientism of era* reasoning — meaning that Le Bon's work was framed
by the 16[10]-19[10] centuries European science. At the time *Psychologie des foules*
was written, the mechanistic paradigm that dominated sciences for more than
three hundred years, was challenged by organismic and holistic paradigms.
The hypothesis of the paper is that Le Bon's concept of the crowd is multiparadigmatic. In testing this hypothesis, it comparatively explores mechanistic,
organismic, and holistic paradigms, and hermeneutically examines their reflection in Le Bon's concept of the crowd.

THE MECHANISTIC PARADIGM

In the 16[10]-18[10] centuries, " the mechanical arts are continually thriving and
growing "[13] and " the Mechanical principles are so universal and therefore
applicable to so many things, they are rather fitted to *include*, than necessitated
to *exclude*, any other hypothesis that is founded in nature "[14]. The primary
mechanistic principle states " the whole is equal to all its parts "[15]. Mechanicism becomes a " ... construction method [*Baukunst*] for putting a world
together out of all kinds of immutable and differently formed material "[16], and
implies " combinatorics "[17]. According to combinatorics, the whole comes to

10. A. Oberschall, *Social Movements : Ideologies, Interests and Identities, op. cit.*, 6-11 ;
W. Gamson, " The Social Psychology of Collective Action ", in A.D. Morris, C.M. Mueller (eds),
Frontiers in Social Movement Theory, New Haven, 1992, 53.

11. R.A. Nye, *The Origins of Crowd Psychology : Gustave Le Bon and the Crisis of Mass
Democracy in the Third Republic*, London, 1975 (SAGE Studies in 20th Century History, vol. 2) ;
S. Barrows, *Distorting Mirrors : Visions of the Crowd in Late 19th-Century France*, New Haven,
1981, 162-188 ; J. van Ginneken, *Crowds, Psychology and Politics 1871-1899*, Cambridge, 1992,
130-187 ; S. Moscovici, *The Age of the Crowd*, Cambridge, [1981] 1985.

12. E.G. King, *Crowd Theory as a Psychology of the Leader and the Led, op. cit.*, 9-41, 42-45,
125-126 ; C. McPhail, *The Myth of the Madding Crowd, op. cit.*, 1-5, 15-23 ; Ch. Lindholm, *Charisma, op. cit.*, 36-49.

13. F. Bacon, *Francis Bacon : A Selection of His Works*, in S. Warhaft (ed.), New York, [1620]
1982, 352.

14. R. Boyle, " About the Excellency and Grounds of the Mechanical Hypothesis ", in M.A.
Stewart (ed.), *Selected Philosophical Papers of Robert Boyle*, Indianapolis, [1674] 1991, 145.

15. E.B. de Condillac, *Philosophical Writings of Etienne Bonnot, Abbé de Condillac*, in F.
Philip (transl.), Hillsdale, 1746-1982, 8.

16. I. Kant, *Opus Postumum*, in E. Förster (ed.), E. Förster, M. Rosen (transl.), Cambridge,
[1801/1936-1938] 1993, 28 (The Cambridge edition of the works of Immanuel Kant). In this passage, which deals with the quantity of matter, Kant uses de term atomism (*Atomistik* see
A. Buchenau, *Kant's Opus postumum*, Zweite Hälfte, Berlin, Leipzig, 1938, 207).

17. I.S. Narskii, *Zapadno-Evropeiskaya filosofia XVII veka*, Moscow, 1977, 17.

existence through " mixture "[18], " arrangement "[19], " aggregation "[20] of inter-dependent parts. Their constitution determines the nature and composition of the whole[21]. Parts themselves are : a) " stable, and not appreciably or perma-nently affected by being part of the system "[22] ; b) " never significantly modi-fied by each other "[23]. The whole does not add anything essential to its parts, and is reduced to the " sum total of its parts "[24], where " *sum* means aggregate without additional principles or regularities "[25]. Therefore, the mechanistic *analytical* method necessitates breaking down complex phenomena into ele-mentary parts and studying them separately. Spinoza writes : " It follows that if a complex object be divided by thought into a number of simple component parts, and if each part be regarded separately[26], all confusion will disap-pear "[27]. This method " presupposes that the splitting of a whole does not affect the character of this whole "[28]. The mechanistic cognition of the whole by the means of " separating out variables "[29] can be reduced to the cognition of its parts and summation of knowledge about them. The reductionism in the construction and functioning of the whole implies what could be called *cogni-tive reductionism*.

The conclusive principle of mechanicism is *mundus omnis, etiam adspect-abilis, machina est*[30].

18. G W.F. Hegel, *Hegel's Science of logic*, A.V. Miller (transl.), London, 1812-1969, 711.

19. E.B. de Condillac, *Philosophical Writings of Etienne Bonnot,...., op. cit.*, 139.

20. E. Montgomery, " The Unity of the Organic Individual ", *Mind*, 5 (1880), 326.

21. I. Kant, " Mechanismus ", in R. Eisler (ed.), *Kant Lexikon : Naschslagewerk zu Kants sämtlichen Schriften, Briefen und handschriflichem Nachlass*, Berlin, 1930, 348 ; F. Bacon, *Francis Bacon : A Selection of His Works, op. cit.*, 382 ; M. Planck, *Acht Vorlesungen über Theoretische Physik : gehalten and der Columbia University in the City of New York im Frühjahr 1909*, Leipzig, 1910, 96-97.

22. W. Buckley, *Sociology and Modern Systems Theory*, Englewood Cliffs, 1967, 46.

23. K. Deutsch, " Mechanism, Organism, and Society : Some Models in Natural and Social Science ", *Philosophy of Science*, 18 (1951), 234.

24. P.A. Angeles, *The Harper Collins Dictionary of Philosophy*, 2nd edition, New York, 1992, 181-182.

25. D.J. Haraway, *Crystals, Fabrics, and Fields : Metaphors of Organicism in 20th-Century Developmental Biology*, New Haven, 1976, 28.

26. L. von Bertalanffy, *Perspectives on General System Theory : Scientific-Philosophical Studies*, in E. Taschdjian (ed.), New York, 1975, 150 ; D. Bohm, *Wholeness and the Implicate Order*, London, 1980, 2-3.

27. B. Spinoza, *Ethics*, Buffalo, [1677] 1989, 21.

28. M. Planck, *Acht Vorlesungen über Theoretiche Physik, op. cit.*, 97.

29. B.G. Hanson, *General Systems Theory Beginning with Wholes*, Washington, 1995, 5.

30. C. Wolff, *Gesammelte Werke*, Herausgegeben und bearbeitet von J. Ecole, J.E. Hofmann, M. Thomann, H.W. Arndt, II. Abteilung - Lateinische Schriften, Band 4 *Cosmologia generalis*, Herausgegeben und bearbeitet von J. Ecole, Hildesheim, [1737] 1964, 67 ; Voltaire, *A Philosophical Dictionary*, vol. 3, London, [1764-1769] 1824, 256 ; A. Smith, " Essays on Philosophical Subjects ", *The Early Writings of Adam Smith*, in J.R. Lindgren (ed.), New York, [1795] 1967, 66 ; J.C. Gottsched, *Erste Gründe der gesamten Weltweisheit*, vol. 1, Leipzig, 1733-1734, 173 ; I. Kant, " Mechanismus ", *op. cit.*, 348.

In the 17th-18th centuries, Descartes[31], Diderot[32], Hobbes[33], and La Mettrie[34] applied the concept of machine to the study of living bodies, viewed as
" mechanical contrivance "[35] or " self-winding machine "[36]. In the 19th century, Prince writes that " mental processes are automatic "[37]. At the beginning
of the 20th century, Ward compares society to a " social machinery "[38]. The
concept of *machine-automaton*, introduced in the 16th-17th centuries[39] pre-
sumes that parts of the whole, resembling a machine, are connected in a unicausal determinism. The unicausal determinism as an integral principle of
mechanicism[40] is based on two facts. First, changes in the whole originate in
a single source of activity or *formative force*[41], namely the prime force, which
is superior to the parts. Second, changes of the parts are one-directed (from the
prime force to the parts) because of their rigid connection to, and dependance
on, the prime force imposing its motion upon them. Mill writes : " In the difficult process of observation and comparison ... it would evidently be a great
assistance if it should happen to be the fact, that some one element in the complex existence of social men is pre-eminent over all others as the prime agent
of social movement. For we could then take the progress of that one element
as a central chain, to each successive link of which, the corresponding links of
all the other progressions being appended "[42]. The idea of unicausal determinism is already contained in the scientific systems of Galileo[43], Descartes[44],
Newton[45], Leibniz[46], and Malebranche[47]. Thus, the concluding paradigmatic

31. R. Olson, *The Emergence of the Social Sciences, 1642-1792*, New York, 1993, 25.

32. O. Mayr, *Authority, Liberty & Automatic Machinery in Early Modern Europe*, Baltimore, 1986, 78.

33. F. Tönnies, *On Social Ideas and Ideologies*, in E.G. Jacoby (ed.), New York, 1974, 35.

34. B. Campbell, " La Mettrie : The Robot and the Automaton ", *Journal of the History of Ideas*, vol. XXXI, n° 4 (1970), 555-572.

35. F. Tönnies, *On Social Ideas and Ideologies*, op. cit., 20.

36. J.O. de La Mettrie, *Man a Machine ; and, Man a Plant*, in R.A. Watson and M. Rybalka (transl.), Indianapolis, [1748] 1994, 32.

37. M. Prince, " The Nature of Mind and Human Automatism ", in W. Bringmann (ed.), *The Origins of Psychology : A Collection of Early Writings*, vol. I, New York, [1885] 1975, 115.

38. L.F. Ward, *Pure Sociology*, New York, 1903, 169.

39. *The Compact Edition of the Oxford Dictionary*, Oxford, 1986, 144.

40. I.V. Blauberg, V.N. Sadovsky, E.G. Yudin, *Systems Theory : Philosophical and Method-ological Problems*, in S. Syrovatkin and O. Germogenova (transl.), Moscow, 1977, 21 ; J. Dupré, *The Disorder of Things : Metaphysical Foundations of the Disunity of Science*, Cambridge, Massachusetts, London, 1993, 2-3 ; J. Battista, " The Holistic Paradigm and General System Theory ", *General Systems*, vol. 22 (1977), 65.

41. I. Kant, *Opus Postumum*, op. cit., 40.

42. J.S. Mill, *A System of Logic*, 8th ed., New York, 1874, 640.

43. M. Bunge, *Causality and Modern Science*, 3rd ed., New York, 1979, 4.

44. See, e.g., S. Gordon, *The History and Philosophy of Social Science*, London, New York, [1991] 1993, 591 ; D. Garber, " Descartes and Occasionalism ", in S. Nadler (ed.), *Causation in Early Modern Philosophy*, Pensylvania, 1993, 10-15.

45. R.H. Hurlbutt, *Hume, Newton, and the Design Argument*, Lincoln, [1963] 1985, 9-16.

46. W. Carr, *Leibniz*, London, 1929, 215 ; I.S. Narskii, *Zapadno-Evropeiskaya filosofia XVII veka*, op. cit., 307.

47. D. Garber, " Descartes and Occasionalism ", op. cit., 25.

principle of mechanicism may be redefined as " the whole is a machine-automaton with a single source of activity ".

THE RISE AND EVOLUTION OF THE ORGANISMIC PARADIGM

In the 19th century the mechanistic paradigm is challenged by the organismic paradigm (organicism), rooted in the ancient philosophers' idea the whole is different from its parts[48], which is reformulated *the whole is more than the sum of its parts*[49]. The intriguing question is what makes the whole greater than the sum of its parts. Yet ancient philosophers direct the answer towards the discovery of the *properties of the whole*, called " the unchanging rules "[50] or " generic differentia "[51]. Montgomery describes them as " fundamental properties of life "[52]. Comte differentiates " the vital phenomena " of " organism " from " psychological laws of the individual " and " actions of individuals on one another "[53]. Spencer discovers an " integrity of the whole " in its " prime traits "[54]. According to Durkheim, " The whole does not equal the sum of its parts ; it is something different, whose properties differ from those displayed by the parts from which it is formed ". He states that " the group thinks, feels and acts entirely differently from the way its members would if they are isolated ", and " society is not mere sum of individuals, but the system formed by their association represents a specific reality which has its own characteristics ". " Own characteristics " of society are, indeed, social facts, existing " outside the consciousness of the individual "[55].

Contemporary scientists define properties of the whole as *emergent properties*[56], which are not reducible to the properties of the sublevels of the system[57]. These properties are engendered by the whole and meaningless in terms

48. Lao-Tzu, *The Tao-Teh-King : Sayings of Lao-Tzu*, in C.S. Medhurst (transl. with commentary), Wheaton, [B.C. 600] 1972, 93 ; Aristotle, *Metaphysics*, Cambridge, 1975, 423.

49. See, e.g., L. Bertalanffy, *Perspectives on General System Theory : Scientific-Philosophical Studies*, op. cit., 149 ; M. Bunge, " General Systems and Holism ", *General Systems : Yearbook of the Society for General Systems Research*, vol. 22 (1977), 87 ; D.C. Phillips, *Holistic thought in Social Science*, Stanford, 1976, 6.

50. Lao-Tzu, *The Tao-Teh-King : Sayings of Lao-Tzu*, op. cit., 50.

51. Aristotle, *On the Parts of Animals*, in W. Ogle (transl. and ed.), New York, 1987, 11.

52. E. Montgomery, " The Unity of Organic Individual ", *op. cit.*, 322.

53. A. Comte, " Plan of this Course, or General Consideration of Hierarchy of Positive Science ", in M. Clark (transl.), *The Essential Comte*, London, [1830-1842] 1974, 53, 56.

54. H. Spencer, *The Principles of Sociology*, New York, 1897, 456.

55. E. Durkheim, *The Rules of Sociological Method and Selected Texts on Sociology and Its Method*, in S. Lukes (ed.), W.D. Halls (transl.), New York, [1895] 1982, 128, 129, 51-52.

56. W.E. Agar, *A Contribution to the Theory of the Living Organism*, Melbourne, [1943] 1951, 39 ; D.C. Phillips, *Holistic thought in Social Science, op. cit.*, 14 ; M. Bunge, " General Systems and Holism ", *op. cit.*, 87 ; A Giddens, *The Constitution of Society : Outline of the Theory of Structuration*, Cambridge, 1984, 171 ; P. Checkland, J. Scholes, *Soft Systems Methodology in Action*, Chichester, 1990, 19.

57. M. Jammer, " Integrative Concepts in the Physical Sciences ", in M. Alonso (ed.), *Organization and Change in Complex Systems*, New York, 1990, 250.

of the parts, and could not be found in them[58]. However, the term *emergent properties* is too ambiguous[59]. Three different groups of emergent properties arise from bringing parts together : 1) emergent properties belonging exclusively to the whole *per se* ; 2) *emergent relational properties*[60] inherent to interactions among parts ; 3) emergent properties of parts caused by their interactions. In our analysis of emergent properties, we focus on the first type, those referring to the properties of the whole *per se*.

Selvin and Hagstrom, largely influenced by Lazarsfeld, distinguish two types of them : 1) " aggregative properties, which are based on characteristics of smaller units within the group " ; 2) " integral properties, which are not based on smaller units "[61]. Remender outlines that " integral properties at each successively higher level of complexity " give us a clue " to the identity of those emergent properties which make each level distinct and irreducible to simpler levels of analysis "[62]. Russian scholars term properties of the whole *integrativnye svoistva* (integrative properties) because they integrate parts into *organic unity*[63]. The idea of integrative properties is central to concepts of *integrative structures*[64], *integrating social associations*[65], *integrated systems*[66], or *integrated wholes*[67]. Thus, it is correct to call properties of the whole *integrative qualities*, with the primary function of system integration.

The integrative qualities may be viewed as dominant and existing prior to the parts and their relations[68]. At this point, organicism evolves into *holism*. The prime focus of holism is " the whole determining the parts " — meaning that " there is nothing in the parts that is not a manifestation of the whole ", and " in all that the parts do and all that they are they only show forth the whole "[69]. According to Durkheim's holistic analysis, society exerts, by the

58. P. Checkland, J. Scholes, *Soft System Methodology in Action, op. cit.*, 18-19.

59. P. Lazarsfeld, H. Menzel, " On Relation Between Individual and Collective Properties ", in A. Etzioni (ed.), *Complex Organizations ; a sociological reader*, New York, 1961, 429.

60. D.C. Phillips, *Philosophy, Science, and Social Inquiry*, Oxford, 1987, 68.

61. H. Selvin, W. Hagstrom, " The Empirical Classification of Formal Groups ", *American Sociological Review*, vol. 28 (1963), 402-403.

62. P. Remender, " Social Facts and Symbolic Interaction : A Search for Key Social Emergent in Durkheim's Sociological Analysis ", in P. Hamilton (ed.), *E. Durkheim : Critical Assessments*, London, [1973] 1990, 347.

63. V. Afanas'ev, *Systemnost' i obshestvo*, Moscow, 1980, 191 ; I.V. Blauberg, V.N. Sadovsky, E.G. Yudin, *Systems Theory : Philosophical and Methodological Problems, op. cit.*, 22 ; V. Kuzmin, *Printsip sistemnosti v teorii i metodologii K. Marksa*, Moscow, 1980, 81-82 ; U. Urmantsev, " Tektologiya i obshaya teoriya system ", *Voprosy Filosofii*, 8 (1995), 18.

64. T. Parsons, *The Social System*, Glencoe, 1951, 137.

65. P.M. Blau, *Inequality and Heterogeneity : a primitive theory of social structure*, New York, 1977, 199.

66. D.C. Phillips, *Philosophy, Science, and Social Inquiry, op. cit.*, 77.

67. P. Lowenhard, " Mind : Mapping the Reconstruction of Reality ", in M. Alonso (ed.), *Organization and Change in Complex Systems*, New York, 1990, 130.

68. E. Montgomery, " The Unity of Organic Individual ", *op. cit.*, 326, 469 ; W. Ogle, " Notes ", in Aristotle, *On the Parts of Animals*, [1882] 1987, 153.

69. J.S. Haldane, " Life and Mechanism ", *Mind*, 9 (1884), 38.

means of social facts, "external constraint", "compelling and coercive power" upon individuals. By virtue of this power, social facts "exercise control over" individuals by "imposing themselves upon them"[70]. Therefore, society and social facts are "prior to any given individual" and "are conceived of as molds which shape the individual's behavior"[71].

In holism, the whole becomes a dominating totality that determines and subordinates the parts[72], and even imposes "likeness"[73] upon them. Evidently, we deal with another form of reductionism, that is *holistic reductionism*, according to which the parts yield their identity to dominating totality of the whole. This reductionism constitutes the core of holistic methodology, which : a) considers the whole to be more powerful and better than its parts[74] ; b) studies the whole unit rather than its parts[75] ; c) denies any influence of isolated parts on the whole[76]. The logical conclusion of *holism*, as James sarcastically deduced, is " if not the Whole System, then Nothing "[77].

MECHANICISM, ORGANICISM, AND HOLISM : COMPARATIVE ANALYSIS

According to mechanicism, the whole is *hysteron, i.e. posterior* (Aristotle) or *a posteriori* (Kant) to its parts. The mechanistic epistemology is "founded on the principle of beginning with the parts and building up the whole by synthesis". The contrary is true in organismic epistemology, where " ... we begin with the whole and arrive at the parts by analysis"[78]. The holistic epistemology states that " a real *whole* necessarily precedes its parts (...) "[79]. The whole is not only *a priori* to its parts but becomes a dominating *archetype*. Durkheim's archetype, *collective conscience*, " completely envelopes our total consciousness, coinciding with it at every point. At that moment our individuality is zero "[80].

70. E. Durkheim, *The Rules of Sociological Method and Selected Texts on Sociology and Its Method*, *op. cit.*, 51-52, 56.

71. P. Remender, " Social Facts and Symbolic Interaction : A Search for Key Social Emergent in Durkheim's Sociological Analysis ", *op. cit.*, 343.

72. W.E. Agar, *A Contribution to the Theory of the Living Organism*, *op. cit.*, 34 ; D.C. Phillips, *Holistic thought in Social Science*, *op. cit.*, 28 ; M. Bunge, " Systems Everywhere ", in C.V. Negoita (ed.), *Cybernetics and Applied Systems*, New York, 1992, 37.

73. E. Montgomery, " The Unity of Organic Individual ", *op. cit.*, 333.

74. W. Gasparski, " System Approach as a Style : A Hermeneutics of Systems ", *Systems Thinking in Europe*, in M.C. Jackson (ed.), New York, 1991, 20.

75. A. Smith, *The concept of social Change : a critique of the functionalist theory of social change*, London, 1973, 27.

76. D.C. Phillips, *Holistic thought in Social Science*, *op. cit.*, 28 ; M. Bunge, " General Systems and Holism ", *op. cit.*, 87.

77. W. James, " Absolutism and Empiricism ", *Mind*, 9 (1884), 283.

78. J.C. Maxwell, *A Treatise on Electricity and Magnetism*, vol. I, Oxford, 1873, xi.

79. I. Kant, *Opus Postumum*, *op. cit.*, 66.

80. E. Durkheim, *The Rules of Sociological Method and Selected Texts on Sociology and Its Method*, *op. cit.*, 84.

The mechanistic mode of thinking either denies the reality of properties of the whole[81], or considers that all properties of a whole may be approached by a study of its parts[82]. The holistic paradigm sees nothing but the whole and its properties. The organismic paradigm analyzes the whole to : a) uncover integrative qualities ; b) distinguish them from qualities of parts ; c) study parts in relation to each other and the whole. The organismic paradigm, contrary to mechanistic and holistic reductionism, implies — reflecting the complexity of the whole — *cognitive comprehensiveness.* Organicism, contrary to holism, neither rejects, nor diminishes the role of the parts and their qualities emerging from interaction with each other. Moreover, it emphasizes *whole-parts-relations* co-determination — something analogous to Bandura's " reciprocal determinism "[83] model, which outlines " reciprocal influence " and " mutual actions between causal factors " which " operate interactively as determinants of each other "[84].

In the organismic paradigm, the wholeness of the system is rested upon triadic co-determination : compatible parts, dynamic interdependence of the parts, and integrative qualities. In the mechanistic paradigm, the wholeness of the system is rested upon dyadic co-determination : compatible parts organized around the prime part (unicause) and rigid/static interdependence of the parts. In the holistic paradigm, the wholeness of the system is unifactorial, as it is founded upon the dominating totality of the whole. Its parts yield their identity to the whole and " disappear " in its totality. The triadic dynamic diversity " whole-parts-relations ", inherent to organicism, are replaced in holism by static sameness of the whole and parts.

Mechanistic and holistic paradigms are modeled upon unicausal determinism. However, each paradigm defines the unicause (prime part, formative force, prime agent) differently. In mechanism, the unicause is either one of the whole's parts, or an outside force ; in holism, the unicause is identical to dominating totality of the whole. The organismic paradigm emphasizes that " there are no absolutely primary parts of matter "[85] and, therefore, stresses multicausality. In a later writing, Durkheim rejected the idea that " in each state of society there is always one element which remains constant, dominating all the others and constituting the prime mover in the progress of society ".

81. J. Agassi, " Methodological Individualism ", *The British Journal of Sociology,* 11 (1960), 244 ; A. Giddens, *The Constitution of Society, op. cit.,* 24.

82. M. Planck, *Acht Vorlesungen über Theoretiche Physik, op. cit.,* 96. (Translation by the author, O.I. Gubin).

83. A. Bandura, " The Self System in Reciprocal Determinism ", *American Psychologist,* 33, n° 4 (1978), 344-358.

84. A. Bandura, *Social Foundations of Thought and Action,* Englewood Cliffs, 1986, 23, 43.

85. I. Kant, *Opus Postumum, op. cit.,* 28.

He rather argued that " there is nothing to justify ... that there is one social phenomenon which enjoys such a prerogative over all the others "[86].

The mechanistic, organismic and holistic paradigms are compatible with
different social systems. Durkheim's organicism informed the liberal demo-
cratic model. Popper ([1945] 1992) links holistic methodology and totalitarian
socio-political order[87]. Mayr's study concludes that mechanicism upholds
determinism, rejects free will, and advocates an " authoritarian conception of
order "[88].

PARADIGM SHIFTS AND COMBINATION IN LE BON'S CONCEPT OF CROWD

Le Bon's analysis of the crowd is accompanied by two paradigm shifts from
organicism to holism and from holism to mechanicism with consequent combination of these paradigms.

a) Implication of organismic paradigm : the crowd-whole is greater than
some of its parts, integrative qualities of the crowd

In Psychologie des Foules, the crowd originally appears as gathering,
agglomeration, aggregate, or combination of heterogeneous elements[89]. This
terminology is borrowed from the mechanistic paradigm. However – as these
terms are defined by the author — they ultimately convey an organismic meaning. Le Bon states that " in the aggregate which constitutes a crowd there is in
no sort a summing-up of or an average struck between its elements ", but rather
takes place " a combination followed by the creation of new characteristics "[90].
The new characteristics are further defined as general characteristics[91]. Le
Bon's description of the crowd's general characteristics as being " peculiar to
crowds ", " very different from those of the individuals composing it " and
" not possessed by isolated individuals[92] "is critical for considering his
approach as organismic in origin.

Le Bon distinguishes two groups of general characteristics. The first group
includes the mental unity and the collective mind, both indicating the emergence of the crowd[93]. The second group includes affirmation, repetition, and

86. E. Durkheim, The Rules of Sociological Method and Selected Texts on Sociology and Its
Method, op. cit., 185.
87. K. Popper, Otkrytoe obshestvo i ego vragi, Moskva, 1992 (Engl. original : The Open Society and its Enemies, vol. 2, London, 1945).
88. O. Mayr, Authority, Liberty & Automatic Machinery in Early Modern Europe, op. cit., 120.
89. G. Le Bon, The Crowd, op. cit., 43, 44, 47.
90. Idem, 47.
91. Idem, 45.
92. Idem, 43, 44, 50.
93. Idem, 44-48.

contagion, which characterize crowd's dynamics and actions[94]. These general characteristics secure unity-wholeness of the crowd, integrate rational individuals in it, and convert them into crowd participants. Thus, Le Bon's general characteristics of the crowd — *the mental unity of the crowd, the collective mind, affirmation, repetition,* and *contagion* — are identical, in terms of the organismic paradigm, with integrative qualities of the crowd.

b) Shift from organismic to holistic analysis : predominant role and coercive power of the integrative qualities, parts give up their individuality and become homogeneous

At the next stage of crowd development, its integrative qualities begin to dominate rational individuals, and *heterogeneous individuals* lose their individuality. Distinctive personalities of *isolated individuals* are paralyzed or demolished by uniform and identical actions of the crowd. In " peculiar acts of the crowd ", " the heterogeneous is swamped by the homogeneous ", " the sentiment of responsibility which always controls individuals disappears entirely ", the individual " yields to instincts which, had he been alone, he would have kept under restrain ", and, finally, " the unconscious qualities obtain the upper hand "[95]. Under the influence of the collective mind and contagion, an " individual readily sacrifices his personal interest to the collective interests "[96]. According to Le Bon, the dominating crowd transforms distinctive personality into an *ordinary individual* who posseses *mediocre qualities*[97]. The individual is reduced by the crowd to a *primitive being* who " descends several rungs in the ladder of civilization ". To display that the crowd-totality completely modifies an individual, Le Bon concludes, " Isolated, he may be a cultivated individual ; in a crowd, he is a barbarian — that is, a creature acting by instincts "[98].

Thus, we have the first paradigm shift from organicism to holism. Le Bon begins his analysis describing the crowd as an organic collectivity with its distinctive integrative qualities that are completely different and cannot be deduced from the rational individuals entering the crowd. His further conceptualization implants following holistic parameters : a) integrative begin to dominate rational individuals ; b) rational individuals lose their individuality, suspended by uniform actions of the crowd, and become mediocre crowd participants ; c) the crowd no more resembles an organism, it becomes a homogeneous totality of mediocre crowd participants whose actions are characterized by rigid sameness and oneness.

94. G. Le Bon, *The Crowd, op. cit.*, 44-54 ; 146-151.
95. *Idem,* 59-60.
96. *Idem,* 50.
97. *Idem,* 48-49.
98. *Idem,* 52.

c) Shift from holism to mechanicism : leader as a unicause, crowd-whole as a machine-automaton, authoritarian order in the crowd

However, Le Bon's further description of the crowd is not holistic. The crowd is described as a hierarchically organized unit — a pyramid with the crowd leader on the top, the mediocre crowd participants at the bottom[99]. The crowd acts like a machine, in which each individual acts as an automaton in unison with the leader. The entire crowd life is organized around a single source of activity, manifested in the will of the leader. The leader's will is " the nucleus around which the opinions of the crowd are grouped and attain unity "[100]. The leader possesses the " persistent will-force ", " an immensely powerful facility to which everything yields ". Le Bon concludes : " Nothing resists it ; neither nature, gods, nor man "[101].

Ultimately, Le Bon's model of crowd action is centered in a rigorous deterministic will of the crowd leader, who facilitates a mental unity of crowd participants by orchestrating contagion. Before contagion, the leader and rational individuals, who enter the crowd, have relatively independent wills and opinions. However, after the contagion is enforced, opinions and will of the leader become predominant, absorb those faculties of rational individuals, and convert them into mediocre crowd participants. Furthermore, the leader hypnotizes crowd participants through repetitive suggestion and affirmation, and his superior will. The contagion, enforced by the crowd's intolerance, conservatism, and autocratic rule, reduces crowd participants to a mere " automaton who has ceased to be guided by his will " and to " a creature acting by instinct "[102]. Thus, the crowd's organization and functioning fits the mechanistic paradigm best because Le Bon : a) describes the crowd in terms of machine-automaton : b) models its actions upon unicausal determinism : c) attributes " authoritarian " and even " despotic authority "[103] to the crowd's leader : d) depicts the crowd as an authoritarian collectivity.

HERMENEUTICAL INTERPRETATION OF LE BON'S CONCEPT OF THE CROWD

Originally founded by Ast[104] and Schleiermacher[105] as an art of interpreting texts[106], hermeneutics was later developed by Gadamer[107] into a scientific

99. G. Le Bon, *The Crowd, op. cit.*, 139-145.

100. *Idem*, 139-140.

101. *Idem*, 144.

102. *Idem*, 52.

103. *Idem*, 74, 142.

104. F. Ast, *Grundlinien der Grammatik, Hermeneutik und Kritik*, Landshut, 1808.

105. F.D.E. Schleiermacher, *Hermeneutik und Kritik*, in M. Frank (ed.), Frankfurt am Main, [1829/1838] 1977.

106. J. Grondin, *Introduction to philosophical Hermeneutics*, New Haven, 1994, 64.

107. G.H. Gadamer, *Truth and Method*, 2nd ed., in J. Weinsheimer and D.G. Marshall (transl.), New York, [1960] 1989.

method of exploring and understanding truth. The " fundamental law " of
hermeneutics states that " all understanding and knowing is to discover the
spirit of the whole in the individual and to grasp the individual in terms of the
whole "[108]. The " whole-individual " relations are dialectical or *dialogical*[109],
resembling the interminable " circular relationship "[110], also known as *hermeneutic circle*[111]. In the hermeneutic circle, " harmonizing " between a whole
and its individual parts is " the criterion of correct understanding. Its absence
means the failure to understand "[112].

This paper presumed that Le Bon's concept of the crowd (*individual*) is constructed within the frameworks of the 16th-19th centuries European science
(*whole*), in which — while Le Bon was developing *Psychologie des Foules* —
organismic and holistic paradigms challenged mechanicism. The dialectical-
dialogical relationship between the 16th-19th centuries European science and
Psychologie des Foules is being threefold. First, the presence of paradigms
(*whole*) in Le Bon's concept of the crowd (*individual*) indicates that the challenge of organicism and holism to mechanicism did not replace " abandon
the " by : lead to total abandonment of mechanistic methodology, but rather
enhanced paradigmatic plurality in the European scientific community. Second, Le Bon's concept of the crowd (*individual*) has become a reflective appro-
priation of *the spirit of the whole* (paradigm plurality in the 16th-17th centuries
European science) and an articulation of this spirit in the part of this spirit. And
third, paradigm combination in Le Bon's concept of the crowd indicates harmony between the 16th-19th centuries European science (*whole*) and *Psychologie des Foules* (*individual*), a deep entrenchment of Le Bon's research into the
science of his time. This entrenchment, represented by paradigm combination
and paradigm shifts in *Psychologie des foules*, informs Le Bon's theoretical
construction of the crowd as a collectivity that transforms *individuals*, the
agents of rationality, morality, civilization, high culture and intellectual
achievement into *masses and crowds*, harbingers of irrationality, destruction,
decadence, the herd instinct, barbarism[113].

Speaking Ast's language, Le Bon eventually grasped one of the *individuals*
of his epoch (the crowd) *in terms of the whole* (multiplicity of paradigms). Le
Bon never admitted the multiparadigmatic nature of his research, but the
hermeneutics let this paper discover — for the sake of " correct under-

108. F. Ast, *Grundlinien der Grammatik, Hermeneutik und Kritik, op. cit.*, 116.

109. J. Bleicher, *The Hermeneutic Imagination : Outline of a Positive Critique of Scientism
and Sociology*, London, 1982, 75.

110. G.H. Gadamer, *Truth and Method, op. cit.*, 291.

111. J. Grondin, *Introduction to philosophical Hermeneutics, op. cit.*, 74.

112. H.G. Gadamer, " On the Circle of Understanding ", in J.M. Connolly, T. Keutner (eds),
Hermeneutics Versus Science : three German views essays, Notre Dame, 1988, 68.

113. A. Oberschall, *Social Movements : Ideologies, Interests and Identities, op. cit.*, 4-5.

standing " (Gadamer) — a multiparadigmatic truth about *Psychologie des Foules* for the author.

OSWALD SPENGLER : SCIENCE, TECHNOLOGY AND THE FATE OF CIVILIZATION

Antonello LA VERGATA

The philosopher Oswald Spengler played an important role in the so-called *Die Streit um die Technik*, the "debate about technology" which came to a head in Weimar Germany and which produced hundreds of books and articles by philosophers, scientists, engineers, politicians, social writers, artists, and industrialists. A non-academic figure, and one who had troubles with the university establishment throughout his life, Spengler discussed technology in the final section of his controversial but enormously successful *Der Untergang des Abendlandes*[1] and in a lecture which he published in 1931 with the title *Der Mensch und die Technik*[2].

Being the most visible aspect of modernity, technology was under concentric fire from neo-Romantics, representatives of the so called conservative revolution, expressionists, *Lebensphilosophen*, existentialists, irrationalist opponents of positivism and scientism, left-wing critics of capitalism, alienation, and reification. Many of these authors equated technological advances with the domination of the spiritless over the spiritual : technology, they said, conquered but de-humanized the world. It stood in opposition to life. It was also the symbol, or the main cause of, in turn, capitalism, *Amerikanismus*, Fordism, mass production, and consumption. A few quotations should be useful in the sketching out of certain aspects of these widespread attitudes.

Technological development, the philosopher and sociologist Georg Simmel feared, might "turn the apparatus into a self-sufficient being " ; " a slave revolt of the means against the ends " would then ensue, " the slave - the means -

1. O. Spengler, *Der Untergang des Abendlandes. Umrisse einer Morphologie der Weltgeschichte*. I. *Gestalt und wirklichkeit*, Wien, Leipzig, 1918 ; II. *Welthistorische Perspektiven*, München, 1922.

2. O. Spengler, *Der Mensch und die Technik : Beitrag zu einer Philosophie des Lebens*, München, 1931.

becoming the master of his master - man "[3]. In an essay of 1931 on *Menschen-fresser Technik*, the journalist Ernst Niekisch described the " antilife [*lebens-feindlich*], demonic quality of technology " in the following words : " Tech-nology is the rape of nature. It brushes nature aside. [...] When technology tri-umphs, nature is violated and desolated. Technology murders life by striking down, step by step, the limits established by nature. It devours men and all that is human "[4].

Even some defenders of technology evince a mixed attitude. The industrial-ist and politician Walther Rathenau, president of AEG (*Allgemeine Elektricitäts-Gesellschaft*), concluded his celebration of the " mechanization of the world "[5] with a couple of ambiguous pages which show a sense of spiritual loss : " Work is no longer a vital occupation, an adaptation of the body and the soul to natural forces ; it has become largely an activity foreign to life's ends, an adaptation of the body and the soul to the mechanical ". Under the deluge of facts, events and information which the modern means of communication inflict upon man, Rathenau went on to say, astonishment, " respect for the event " and human sensibility all dwindle. The senses must be ever more ener-getically stimulated. If this does not occur, despondency and boredom poison life. But if hyperstimulation does occur, a dramatic escalation begins : the delight in nature and art yields to sensational, hasty, false, and poisoned plea-sures, which border on despair, and may eventually lead to it. One is reminded, Rathenau added, of the Pretenders in Ulysses' house, " who kept laughing and eating bleeding meat, while their cheeks were furrowed by tears " . As an instance of the degenerate enjoyment of nature Rathenau cited a car drive for miles and miles ; as an instance of perverted artistic sensibility he mentioned the whodunnit film. In all these follies, as in all excessive forms of stimulation, there was, according to him, " something mechanical ". Rathenau also occa-sionally joined the chorus of his fellow countrymen's *Antiamerikanismus* : " America ", he once confessed to André Gide, " has no soul, and doesn't deserve one, for it has not yet accepted to plunge into the abyss of suffering and sin "[6].

In a very important book on what he defined as *Reactionary modernism*[7], Jeffrey Herf has shown that in those very 1920s and 1930s there was produced in Germany a synthesis of hostility towards industrial society and fascination

3. G. Simmel, *Philosophie des Geldes,* Leipzig, 1900, 521-522, Idem, *Der Konflikt der mod-ernen Kultur. Ein Vortrag*, Leipzig, 1918.

4. E. Niekisch, " Menschenfresser Technik ", *Widerstand*, 6 (1931), 110 ; Engl. transl. of this passage in J. Herf, *Reactionary Modernism. : Technology, culture, and politics in Weimar and the Third Reich*, Cambridge, 1984, 39 (rept. 1987).

5. W. Rathenau, " Die Mechanisierung der Welt ", *Zur Kritik der Zeit*, Berlin, 1912, 45-95, 94.

6. A. Gide, *Journal (1889-1939)*, Paris, 1948, 713 ; quoted in T. Maldonado (ed.), *Tecnica e cultura. Il dibattito tedesco fra Bismarck e Weimar*, Milano, 1979, 19.

7. J. Herf, *Reactionary Modernism. : Technology, culture, and politics in Weimar and the Third Reich*, *op. cit.*, I am deeply indebted to this important book.

for technological advances. Technology was made compatible with the German revolt against Western capitalism, liberalism, and positivism. The reactionary modernists showed that technology could be legitimated without succumbing to Enlightenment rationality. To recall the keywords of this debate among reactionary modernists : technology could be rescued from *Zivilisation* and reconciled with *Kultur*. This new view of technology was couched in the very language of instinct, blood, organism, nation, soil, spirit, will, duty, life which *literati*, neo-romantics, anti-capitalist *Volk* ideologues had used to oppose it and to appeal to German *Innerlichkeit*.

The ways in which technology could be absorbed by the antimodernist tradition can be summarized as follows :

1) Technology must be placed under the control and in the service of the *Volk* and its will to power. It can thus be redeemed by subordinating it to a racial and religious ethics (an argument later used by the Nazis).

2) It could be argued that the roots of the modern crisis lay not in technology as such, but in the unprecedented domination of the economy over cultural life. It would therefore be a mistake to blame technology for sins whose origin was to be found in the economy. It was the latter, thus, which had stripped nature of its beauty and reduced spiritual creations to the status of commodities. Capitalism was alien to a German tradition within which technology was tied not to the mass production of goods for consumption, but rather was art, liberty, and idealism. German was the only place where technology and spirit could be reconciled.

3) Technology was not dead or soulless. As Carl Schmitt put it in 1932 : " The spirit of technicism that had led to the mass adulation or an antireligious, this-worldly activism is spirit [*Geist*], perhaps an evil and satanic spirit, but not to be dismissed as mechanistic and not to be attributed to technology. It is perhaps something terrifying, but itself nothing technical nor mechanical. It is the belief in an activistic metaphysic, the belief of the limitless power and domination of man over nature, even over the human body, in the unlimited 'recession of natural boundaries', in unlimited possibilities for transforming the naturally constituted human existence. One can call this fantastic and demonic, but not simply dead, spiritless, mechanized soullessness "[8].

The spirit of technicism possessed an elective affinity with Schmitt's authoritarian and decisionist view of politics and his ethic of battle, will and struggle. It enhanced the domination over both nature and human beings. The Romantics' hostility to technology, Schmitt had maintained already in 1919, was due to their " escapist subjectivism " and " unmanly passivity "[9].

8. C. Schmitt, *Der Begriff des Politischen : Mit einer Rede über das Zeitalter der Neutralisierung und Entpolitisierungen*, München, 1932, 79-80, quoted from J. Herf, *Reactionary Modernism. : Technology, culture, and politics in Weimar and the Third Reich, op. cit.*, 120.

9. C. Schmitt, *Politische Romantik*, München, Leipzig, 1919, 88-89, 90, 95, 100.

4) Some reactionary modernists said that the fact that machines had a " life of their own " was not evidence of a curse. Quite the contrary, it showed that they drew on the same pool of primordial forces that man drew on. Far from being inimical to life, technology was rooted in it, or, to be more exact, it was rooted in human instinct, and above all in the " will to power ". Technology, wrote the engineer Eugen Diesel in 1926, could not be confused with Americanism, that is with " the obsession with economy " and the repression of man's instincts as a result of a civilization based upon consumption. The development of technology carried hints of a " nobler race [...] of stronger life instincts "[10]. Technology, wrote the philosopher and sociologist Hans Freyer in 1929, is not a mere set of instruments, but the result of a particular cultural context and the " armament of a particular will ", the objectification of a path that a part of humanity has chosen. It is, therefore, not " neutral with regard to values ", but rather " has its own values as the embodiment of a historical will " : " it is linked to the inner fate of a whole culture "[11]. Far from being a product of capitalism and materialism, it was born before capitalism ; indeed, it could be reconciled with " the revolution of the *Volk* against industrial society ". Modern technology does not pose a threat to the soul[12]. There is no danger in the domination of nature by the machine, no danger that — as Simmel feared — technical development may " turn the apparatus into a self-sufficient being ".

5) Finally, it could be argued that the domination over nature, to which technology contributed, was an enterprise belonging to the realm of *Kultur*, not *Zivilisation*. It is here that a crucial role was played by Spengler.

Spengler's view of technology was integral with his view of science, which, in turn, was in tune with the irrationalism of the period[13]. To this he added an extreme historical determinism and cultural pessimism. Both science and technology were, according to him, imbued with the " Faustian " spirit typical of Western civilization. They displayed the " will to power over nature " which characterize the Faustian man, an insatiable and tragic creature whose craving for domination over the world is a manifestation of a never-to-be satisfied longing for infinity. Science and technology are rooted in an inner, ultimately religious urge. Unlike the " Apollonian " science of Antiquity, which was contemplative and inspired a passive attitude towards nature, Western science has been from the beginning oriented towards practical considerations. Theories

10. E. Diesel, *Der Weg durch das Wirrsal : Das Erlebnis unserer Zeit*, Dritte auflage, Stuttgart, Berlin, 1930, 260-261, quoted from J. Herf, *Reactionary Modernism. : Technology, culture, and politics in Weimar and the Third Reich, op. cit.*, 163.

11. H. Freyer, " Zur Philosophie der Technik ", *Blätter für Deutsche Philosophie : Zeitschrift der Deutschen Philosophischen Gesellschaft*, Band 3 (1929-1930), 192-201.

12. H. Freyer, *Revolution von Rechts*, Jena, 1931, 72.

13. O. Spengler, *Der Untergang des Abendlandes. Umrisse einer Morphologie der Weltgeschichte, op. cit.*

are instruments for mastering nature, not for unveiling the mysteries of the world. Their very contents are shaped by the unconscious drive towards possession : this is why they tend to be formulated in the dead language of number and measurement. The Faustian imperative, that is the essence of modern man, leads in the end to self-destruction, but *all* civilizations, irrespective of whether or not they are technologically advanced, are doomed to extinction. Technology, therefore, is merely an instrument of the fate of our civilization. A heroic and tragic vision was Spengler's way of reconciling life and technology.

Spengler *did* despise many aspects of modernity, and he counterposed *Kultur* to *Zivilisation ;* however, unlike many critics of modernity, he did not see in it the essence of mechanical *Zivilisation* and the negation of life. Technology, he argued, does not originate in an arbitrary act ; it is the result of an immanent necessity, of a destiny. Far from being the antithesis of life, or the consequence of a " loss of soul " *(Entseelung),* it is rooted in life itself, of which it is an emanation. No privilege of man, it is as ancient as animal life. Nor does it depend on the use of instruments : " It is the tactics of the whole of life, the inner form of the method of the struggle for life " (by which Spengler meant the exertion of Nietzsche's will to power rather than the ecological context of Darwin's natural selection). It is verbal language that differentiates human technology from that of animals, by adding to it the element of theory and by bringing it into the sphere of consciousness and the intellect. Thus, on the one hand, technology becomes capable of overcoming the limits of the animal mind, but on the other hand, it comes to suffer from the very shortcomings which Spengler, like other *Lebensphilosophen,* attributes to the intellect as opposed to intuitive insight : it can use, transform, dominate things, but not unveil their mystery ; it has forever ceased to be in the service of life. The modern magician (the technician, the inventor, the scientist), is powerful but ignorant. He cannot help being so. Ignorance is his destiny.

Each *Kultur* has its own technology, in which it makes manifest its soul. Classical man — Euclidean, contemplative and passive towards nature — could hardly understand technology's promise. But modern, Faustian man, with his impulse to transcend finitude, and his craving for power, finds in it the most adequate means of expressing his soul. All Western science has since the beginning been " the organ of the technological will to power ". The enterprise has been so successful that " the machine has dethroned God ". But it has also taken on a life of its own and has enslaved man. No return to primitivism or pastoral reconciliation with nature is any longer possible. The " will to power over nature " compels man to fight the battle to the end. He will be defeated ; he cannot escape his fate. Only the engineer, whom Spengler does not see as a soulless technician but as the crowning embodiment of the Faustian soul, can ride the tiger. But this " silent priest of nature " is, in turn, under the attack of the capitalist. Money will eventually triumph over the machine. Then, another struggle will take place, the final one between money and blood. Blood will triumph, with the aid of the sword. Politics, in the form of Caesarism, will defeat economic reason. This will mark a return in human affairs to pure expressions of barbaric instinct and of the will to power, that is of life. The life cycle of our civilization will end in " the elementary fact of blood, which merge into the eternally circulating cosmic wave ", whence everything comes. In this mixture of cultural pessimism, fatalism, irrationalism and mythology, technology as such is said to be doomed, but it is not condemned as responsible for the end of our civilization.

In *Der Mensch und die Technik*[14], Spengler introduced some alterations in the plot and the characters of this tragedy. He described man — not only Faustian man — as a beast of prey : strong, lonely, unbridled, rebellious, brave, cunning and pityless — an indomitable dominator. He also accentuated the racist and the Nietzschean elements in his theory. He saw the nordic races as possessing true Faustian natures ; he accepted Nietzsche's distinction between lords and slaves, and he further subdivided lordly natures - this too echoing Nietzsche - into priests and warriors (or " Vikings of the spirit " and " Vikings of the blood "). The figure of the engineer somewhat receded in the background, and no mention was made of the struggle between money and blood. At the centre of the stage there were now the *Führernaturen*, the leaders, whom Spengler imagined as a mixture of a chieftain, an entrepreneur, and an inventor. But to an even greater degree than in the Untergang, over the whole of man's history there still hung a sense of tragedy.

Man had abandoned his original union with nature. World history was " the story of an inexorable, fatal separation between the human world and the universe of worlds, the story of a rebel who severed his link with his mother's womb and lifted his hand against her ". But nature is stronger. All civilizations are sentenced to death from the beginning. And the history of the world is but a succession of failures. Nature's revenge takes the form of the inner necessity by which civilizations begin their downward path to self-destruction at the very moment they reach their apex. Driven by his compulsion to dominate, man extends his domination through technology until the whole of civilization becomes one great machine. Then the world society becomes a cage in which the predator himself gets trapped. Nature takes her revenge through the hyperdevelopment of civilization and the hyperorganization of society. No sooner has the mechanization of the world been achieved, than a disintegration from within begins. This process is announced by many symptoms : *taedium vitae*, lassitude, a " hellenistic " love for luxury and pleasure, intellectualism, growing tensions between leaders and masses, diffusion of technology among " non-Faustian races " (African, Asian, Mediterranean), which are interested

14. O. Spengler, *Der Mensch und die Technik : Beitrag zu einer Philosophie des Lebens, op. cit.*

only in using it as a weapon against Western civilization. It is interesting to note that one of these symptoms is the temptation to embrace primitivism. Far from figuring as an escapist movement or as a means of regeneration, primitivism belongs to the terminal phase of society's development ; it is the last delusion before death. One cannot escape one's fate ; one must submit to it, and accept it manfully, like the Roman soldier who was on sentry duty in Pompeii and who, not having been relieved of his service, remained in his place until he died. Or, one might guess Spengler would as well have said, like a true German soldier, loyal and disciplined even in the greatest defeat.

In spite of the alterations, Spengler's message remained the same as in the *Untergang* : man cannot but obey his *Bestimmung*, which is assigned to him by the soul of his *Kultur*. As to technology, it is doomed to death not because it is against life, but, on the contrary, because it obeys life's first command : grow and die. It is the destiny of everything to return to the dark abyss from which it originated. Technology does not necessarily bring disenchantment and rationalization with it. Today engineers and technicians tend to be proud of engaging in the most rational of professions, the one whose essence consists in the utilitarian balancing of means and ends. But in Weimar Germany a paradoxical reconciliation of technology and irrationalism became possible : most engineers and technicians were just as irrational in their outlook as irrationalist philosophers and critics of modernity. Spengler was right, although for the wrong reasons, when he said that the power of forming mythologies is not lost in modern civilization. We often tend to reason and behave as if the technological developments which lie ahead were, by their very nature, both rational and inevitable, as if they were inscribed in the nature of things. But those who endorse mythological concepts such as " the nature of things " or " spontaneous development " (which are equivalent to Spengler's " destiny "), should be aware that this amounts to endorsing the equally mythical idea that there is, as it were, a soul which determines all development. And this is a belief which hardly matches with the dogmatic confidence in the rationality of technology as such.

ZUR GESCHICHTE DER ZEITSCHRIFT " ERKENNTNIS " (1930-1940) IM LICHTE DES BRIEFWECHSELS VON HANS REICHENBACH[1]

Hannelore BERNHARDT

Wissenschaftsorganisatorische Arbeiten sind eine verdienstvolle Tätigkeit vieler Gelehrter und ein nicht zu unterschätzender Beitrag für die Entwicklung der Wissenschaft, einzelner ihrer Gebiete und über längere Zeiträume hinweg auch für die scientific community, die in besonderem Maße mit speziellen Aspekten der Wissenschaftsentwicklung und -geschichte verbunden ist, die sie fördert, manchmal auch hemmt, immer aber in Bewegung hält. In diesem Sinne beanspruchen Gründung, Herausgabe und das weitere Schicksal wissenschaftlicher Zeitschriften einen festen Platz in der Historiographie der Wissenschaften. Hans Reichenbachs Bemühungen um Zeitschriftengründungen erweisen sich so gesehen neben seiner reichen Lehr-und Forschungsarbeit ohne Zweifel sowohl für ihn und seine Beziehungen zu Zeitgenossen mit ähnlichen wissenschaftlichen Vorstellungen und Ambitionen als auch für die Entwicklung bestimmter Bereiche der Philosophie des 20. Jahrhunderts als von großer Bedeutung.

Intentionen Reichenbachs, eine Zeitschrift zu gründen, gehen in die frühen zwanziger Jahre zurück und standen in unmittelbarem Zusammenhang mit der Herausbildung einer für diese Zeit vornehmlich in den Wissenschaftszentren Wien, Berlin und Prag entwickelten eigenen philosophischen Richtung des logischen Empirismus bzw. der wissenschaftlichen Philosophie. Rudolf Carnap erinnerte sich im Jahre 1963 rückblickend : *Among all those who worked in Germany in a similar direction in philosophy and in the foundations of science, Hans Reichenbach was the one whose philosophical outlook was nearest to mine. ... Both of us came from physics and had the same interest in its philosophical foundations, and especially in the methodological problems created by Einstein's theory of relativity. Furthermore, we had a common*

1. Diese Arbeit wurde durch ein Forschungsprogramm der Deutschen Forschungsgemeinschaft gefördert.

interest in the theory of knowledge and in logic. At first we communicated only by correspondence. It was not until March 1923, that we met at a small conference in Erlangen, which we organized with a few others who were likewise working in the field of symbolic logic and its use for the development of a scientific philosophy. Among the participants were Heinrich Behmann, Paul Hertz, and Kurt Lewin. There were adresses on pure logic, e.g., a new symbolism, the decision problem, relational structures, and on applied logic, e.g., the relation between physical objects and sense-data, a theory of knowledge without metaphysics, a comparative theory of sciences, the topology of time, and the use of the axiomatic method in physics. Our points of view were often quite divergent, and the debates were very vivid and sometimes heated. Nevertheless, there was a common basic attitude and the common aim of developing a sound and exact method in philosophy[2].

Aus historischer Sicht war der Plan, eine eigene Zeitschrift zu gründen, die die verstreut erscheinenden Publikationen zur " exakten Richtung in der Philosophie " vereinigen sollte, zwar verständlich, aber unter den Bedingungen fortschreitender Inflation und darüber hinaus auch konzeptioneller Differenzen der Beteiligten zum Scheitern verurteilt gewesen. 1929/1930 aber bestand eine andere Situation.

" Aus den *Annalen der Philosophie* ist in den Händen von Rudolf Carnap und Hans Reichenbach eine neue philosophische Zeitschrift geworden, die nach dem, was bis jetzt von ihr vorliegt, den schönen Namen *Erkenntnis* in einem doppelten Sinne verdient : einmal, insofern sie fast ausschließlich Beiträge zur Analysis von Erkenntnisproblemen im strengsten Sinne des Wortes enthält, sodann, insofern die Methode dieser Analysis der Methode der strengen Wissenschaften musterhaft nachgebildet ist "[3].

Der erste Band der neuen, so gelobten und im weiteren in Fachkreisen anerkannten Zeitschrift war im Auftrage der Gesellschaft für empirische Philosophie Berlin (Berliner Gruppe) und des Vereins Ernst Mach Wien auf intensives Betreiben vor allem Hans Reichenbachs, eines der beiden Herausgeber, im Jahre 1930 der wissenschaftlichen Öffentlichkeit vorgelegt worden[4].

Der in den letzten Jahren bearbeitete Briefwechsel Reichenbachs beleuchtet in spezifischer Weise sowohl die im Vorfeld dieser Gründung geführten, durchaus noch schwierigen Verhandlungen als auch die letztlich erfolglosen Versuche einer Fortsetzung der Zeitschrift in den Jahren nach 1933 bzw. 1937. Die

---

2. R. Carnap, " Intellectual Autobiography " in P.A. Schilpp (Hrsg.), *The Philosophy of Rudolf Carnap*, London, 1963, 14 (The Library of living philosophers, vol. XI).

3. H. Scholz, *Deutsche Literaturzeitung, Wochenschrift für Kritik der Internationalen Wissenschaft*, 52, Band der ganzen Reihe (1931), 1835-1841, 1835.

4. Vertreter des Wiener Kreises waren u.a. R. Carnap, M. Schlick, O. Neurath, H. Hahn, E. Zilsel ; zur Berliner Gruppe zählten u.a. H. Reichenbach, W. Dubislav, K. Grelling, A. Herzberg, W. Köhler, C.G. Hempel.

nachfolgenden Ausführungen werden hauptsächlich den ersten Zeitraum behandeln.

Auf der ersten Tagung für Erkenntnislehre der exakten Wissenschaften im September 1929 in Prag fanden im Zusammenhang mit dem weiteren Schicksal der oben erwähnten Zeitschrift *Annalen der Philosophie*, die sichhinsichtlich Qualität und Zahl der Leser ungünstig entwickelt hatte, Gespräche zwischen dem Leipziger Verleger der Annalen, Felix Meiner, und Vertretern des Wiener Kreises und der Berliner Gruppe über ein Publikationsorgan statt geplant war ein Publikationsorgan für die van den letzteren seit den 20er Jahren vertretene Bewegung des logischen Empirismus, die stärkere, vor allem internationale Beachtung anstrebte und nicht zuletzt dank der gegründeten Zeitschrift auch erlangte, im übrigen auch dank mehrerer internationaler Kongresse in den 30er Jahren.

Am 18. Januar 1930 fragte Verleger Meiner brieflich bei Reichenbach an, ob er " geneigt wäre, die Herausgabe der neuen Zeitschrift zu übernehmen ". Er wolle es " mit Personen zu tun haben, nicht mit Vereinen ", wie offensichtlich zunächst vorgesehen war[5]. Sodann schlug Meiner aus inhaltlichen Gründen vor, der Zeitschrift einen neuen Namen zu geben, doch stünde nichts im Wege zu sagen " zugleich Bd. IX der *Annalen der Philosophie* ". Das programmatische Bekenntnis der Vertreter des Wiener Kreises zur " wissenschaftlichen Weltauffassung " — unter dieser Bezeichnung erschien seit 1929 eine eigene Schriftenreihe — schien dem Verleger allerdings " nicht ganz zweifelsfrei ", was wohl meint : zu einseitig. Er wolle " eine wirklich umfassende philosophische Zeitschrift " haben, fügte aber, an Reichenbach gerichtet, hinzu, da " die Titelfrage von einer genauen Kenntnis der philosophischen Strömungen der Gegenwart abhängt, so möchte ich nicht endgültig entscheiden, sondern muß ich mich ... auf Ihr Urteil verlassen "[6].

Mit Datum vom 4. Februar 1930 schrieb Reichenbach an seinen Kollegen und späteren Mitherausgeber der geplanten Zeitschrift, Carnap : " Der von Neurath ... vorgeschlagene Titel *Die Wissenschaft* ist zu farblos. ... Da ich weiß, daß Ihnen das Wort Naturphilosophie nicht so sympathisch ist, habe ich an den Titel *Erkenntnis* gedacht"[7]. Wenig später liest man an Meiner gerichtet : " Leider stößt dieser Titel bei vielen meiner hiesigen Freunde auf Ablehnung, ... er klänge sentimental oder auch mystisch "[8]. In der Folge gab es eine Reihe von Titelvorschlägen, begründet von der einen Seite und wieder verworfen von

5. HR-025-02-53 F. Meiner an H. Reichenbach am 18, 1, 1930. Zitiert mit Genehmigung der University of Pittsburgh. Alle Rechte vorbehalten.
6. HR-025-02-33 F. Meiner an H. Reichenbach am 6, 3, 1930. Zitiert mit Genehmigung der University of Pittsburgh. Alle Rechte vorbehalten.
7. HR-014-23-03 H. Reichenbach an R. Carnap am 4, 2, 1930. Zitiert mit Genehmigung der University of Pittsburgh. Alle Rechte vorbehalten.
8. HR-025- 02-34 H. Reichenbach an F. Meiner am 4, 3, 1930. Zitiert mit Genehmigung der University of Pittsburgh. Alle Rechte vorbehalten.

der anderen : Naturphilosophie, Einheitswissenschaft, Fundamenta : Zeitschrift für Grundlagenforschung, Philosophische Zeitschrift, Die neue Philosophie, Naturerkenntnis. Nach vielseitig abwägenden Überlegungen einigte man sich schließlich doch auf den Titel *Erkenntnis*.

Hinter solcherart Debatten verbergen sich konzeptionelle Differenzen, die noch weit schärfer als in Überlegungen zur Benennung der Zeitschrift in der Frage nach Stoffgebieten, Inhalt und Methoden der aufzunehmenden Beiträge hervortraten, die gemeinhin leicht verständlich, aber auch gelegentlich anspruchsvoll sein sollten. Carnap vertrat die Meinung, daß als Stoffgebiet keinesfalls " das Gesamtgebiet der Philosophie ", sondern " vielleicht ... die philosophischen (logischen und erkenntnistheoretischen) Grundlagen der Wissenschaften, insbesondere der Naturwissenschaften und der Mathematik " benannt werden sollte[9], wogegen wiederum der Verleger wegen der " Einseitigkeit und Beschränkung ausschließlich auf logische und erkenntnistheoretische Fragen " Bedenken äußerte[10]. Zustimmend erklärte Reichenbach mit dem ihm eigenen großen Selbstbewußtsein, daß " der Terminus 'Gesamtgebiet der Philosophie' nicht die geringste Gefahr " bedeute, " da wir Arbeiten einer von uns abgelehnten Richtung immer als in der Qualität ungenügend bezeichnen könne "[11]. Man solle im übrigen erst einmal anfangen.

Wir übergehen die brieflichen Diskussionen um die letztlich doch nicht erfolgte Konstituierung eines Redaktionskollegiums, die Auswahl der Gutachter und zu juristischen Einzelheiten zwischen den Wissenschaftlern, den Herausgebern der *Annalen* und dem Verlag Meiner.

Die prinzipiellen philosophischen Meinungsverschiedenheiten zwischen Wien und Berlin wurden noch sichtbarer, als Reichenbach seinen " Eröffnungsbeitrag " mit der Überschrift " Zur Einführung " vorlegte. Seine programmatischen Überlegungen bestimmten die Philosophie " ... im Sinne von Wissenschaftskritik zu treiben und durch wissenschaftsanalytische Methoden diejenigen Einsichten in Sinn und Bedeutung menschlicher Erkenntnis zu gewinnen, welche die in immer neuen Systemen formulierte, auf ein angenommenes Eigenrecht der Vernunft gegründete Philosophie der historischen Schulen vergeblich gesucht hatte "[12]. Reichenbach gab der Hoffnung Ausdruck, daß " die Philosophie als Wissenschaft eine neue Grundlegung erfahren wird "[13].

---

9. HR-013-41-69 R. Carnap an F. Meiner am 26, 4, 1930. Zitiert mit Genehmigung der University of Pittsburgh. Alle Rechte vorbehalten.

10. HR-013-24-62 F. Meiner an H. Reichenbach am 29, 4, 1930. Zitiert mit Genehmigung der University of Pittsburgh. Alle Rechte vorbehalten.

11. HR 013-41-63 H. Reichenbach an R. Carnap am 16, 5, 1930. Zitiert mit Genehmigung der University of Pittsburgh. Alle Rechte vorbehalten.

12. R. Carnap, H. Reichenbach, " Zur Einführung ", *Erkenntnis : Im Auftrage der Gesellschaft für Empirische Philosophie Berlin und des Vereins Ernst Mach in Wien*, Erster Band (1930-1931), 1 (Zugleich *Annalen der Philosophie*, Band IX).

13. a. a. O., 2.

" Klarheit der Sprache, Einsicht in die Bedeutung des eigenen Wortes erscheint uns ... als höchstes Erfordernis philosophischen Schrifttums ; ... "[14].

Die eingereichten Arbeiten sollten " ihre Quellen in dem ertragreichen Boden der Empirie "[15] haben. Man wolle keine Lehrmeinungen, keine ausgedachten Systeme. Da Erkenntnis in der Philosophie vergleichbar der in den Einzelwissenschaften anzustrebenden sei, " haben wir das Wort als Zeichen für die neue Zeitschrift gewählt "[16].

Diesen Auffassungen konnten sich weder Carnap noch Schlick anschließen, so daß Reichenbach die " Einführung " allein unterzeichnete. Schlick trat sogar von der vorgesehenen Mitherausgeberschaft zurück und verfaßte einen eigenen Beitrag, " um meiner Sympathie und meinem guten Willen Ausdruck zu geben "[17]. Diese Arbeit erschien ebenfalls in Band 1 der " Erkenntnis " und läßt unterschiedliche Standpunkte zwischen den Berliner und Wiener Vertretern des logischen Empirismus scharf hervortreten. Während von Reichenbach die Philosophie also als Wissenschaft aufgefaßt wurde[18], vertrat Schlick die Auffassung, daß " jede Wissenschaft ... ein System von Erkenntnissen, d.h. von wahren Erfahrungssätzen ; und die Gesamtheit der Wissenschaften, das System der Erkenntnisse " ist, " es gibt nicht außerhalb seiner noch ein Gebiet 'philosophischer' Wahrheiten, die Philosophie ist nicht ein System von Sätzen, sie ist keine Wissenschaft "[19]. " Was ist sie aber dann ? " stellte Schlick selbst die berechtigte Frage und antwortete im Anschluß an Wittgenstein : Philosophie ist Tätigkeit, und zwar jene, " durch welche der Sinn der Aussagen festgestellt oder aufgedeckt wird. Durch die Philosophie werden Sätze geklärt, durch die Wissenschaften verifiziert "[20].

In einer brieflichen Mitteilung Carnaps an Reichenbach zu dessen Entwurf des Eröffnungsaufsatzes heißt es : " Die inhaltlichen Differenzen sind vor allem die folgenden. Sie machen der traditionellen Philosophie hier in einem Grade Zugeständnisse, der mich ... sehr verwundert hat. Wir und auch Sie sind doch in Wirklichkeit nicht der Meinung, daß die Ansichten der verschiedenen philosophischen Systeme untereinander und gegen uns nur darauf beruhen, daß sie sagen, was man noch nicht weiß. Es müßte doch gesagt werden, daß jene Systeme zum großen Teil Metaphysik enthalten, die wir für sinnlos halten. Und wenn Sie von der Philosophie als einer Wissenschaft sprechen, so widers-

14. a. a. O., 3.

15. a. a. O., 1.

16. a. a. O., 3.

17. HR-013-30-28 M. Schlick an H. Reichenbach am 8, 6, 1930. Zitiert mit Genehmigung der University of Pittsburgh. Alle Rechte vorbehalten.

18. H. Reichenbach, " Zur Einführung ", *op. cit.*, 2.

19. M. Schlick, " Die Wende der Philosophie ", *Erkenntnis : Im Auftrage der Gesellschaft für Empirische Philosophie Berlin und des Vereins Ernst Mach in Wien*, Erster Band (1930-1931), 7-8 (Zugleich *Annalen der Philosophie*, Band IX).

20. a. a. O., 8.

pricht das beinahe wörtlich dem, was wir in der Broschüre " Wissenschaftliche Weltauffassung "[21] ... gesagt haben "[22]. Zugleich stellte Carnap als " neuen Kurs dieser Zeitschrift " die Aufgabe, " die neue wissenschaftliche Methode des Philosophierens zu fördcrn ", die " in der logischen Analyse der Sätze und Begriffe der empirischen Wissenschaft besteht "[23]. Reichenbach seinerseits fand die Auffassung, daß Philosophie keine Wissenschaft, sondern eine Tätigkeit sei, " unter logischen Gesichtspunkten sehr interessant " und regte eine " ausführliche Diskussion ... in unserer Zeitschrift " dazu an[24].

Noch im Jahr 1930 sandte Reichenbach Briefe mit der Bitte um Beiträge für die neue Zeitschrift an A. Einstein, B. Russell, E. Cassirer, C.L. Lewis von der Harvard University und andere Adressaten, nicht immer mit positivem Resultat. Beispielsweise hat Einstein niemals in der Erkenntnis publiziert. In der Folgezeit entwickelte sich die Erkenntnis jedoch trotz oder vielleicht gerade wegen zahlreicher unterschiedlicher Ideen und Auffassungen zu einem geschätzten Publikationsorgan, das rasch mehr als 500 Abonnenten im In- und Ausland zählte.

Aber dieser Aufstieg währte nicht lange. Im Jahre 1933 mußte Reichenbach Deutschland verlassen, er ging nach Istanbul und später nach Los Angeles. Carnap siedelte 1936 in die USA über. Ein Großteil der redaktionellen Arbeiten erledigte etwa ab 1934 der in die Niederlande emigrierte Otto Neurath, der damit auch versuchte, einen gewissen Zusammenhalt der Vertreter des logischen Empirismus zu bewirken. Nach dem 3. Internationalen Kongreß für die Einheit der Wissenschaft vom 29. - 31. Juli 1937 in Paris, auf dem u.a. die Weiterführung der Erkenntnis beraten und empfohlen wurde, werden im Briefwechsel nun vor allem die technischen Details des Fortbestandes der Zeitschrift, alsbald unter den Bedingungen der faschistischen " Kulturgesetzgebung ", widergespiegelt.

Bis zum Jahre 1936 konnte die Erkenntnis " unauffällig " in Deutschland weiter erscheinen. Danach geriet auch der Meiner-Verlag wegen der jüdischen Abstammung Reichenbachs und vieler jüdischer Autoren und Abonnenten und wegen des Vorwurfs, daß " die Zeitschrift deutschfeindlich eingestellt sei "[25], in Bedrängnis. Der Verein Ernst Mach war bereits am 23. Februar 1934 von der Dollfuß-Regierung in Wien verboten worden. Nach langen Verhandlungen

21. O. Neurath, *Wissenschaftliche Weltauffassung. Der Wiener Kreis*, (zus. mit H. Hahn und R. Carnap), Wien, 1929 (Veröffenticungen des Vereines Ernst Mach).

22. HR-013-41-66 R. Carnap an H. Reichenbach am 29, 4, 1930. Zitiert mit Genehmigung der University of Pittsburgh. Alle Rechte vorbehalten.

23. R. Carnap, " Die alte und die neue Logik ", *Erkenntnis : Im Auftrage der Gesellschaft für Empirische Philosophie Berlin und des Vereins Ernst Mach in Wien*, Erster Band (1930-1931), 12 (Zugleich *Annalen der Philosophie*, Band IX).

24. HR-013-41-65 H. Reichenbach an R. Carnap am 6, 5, 1930. Zitiert mit Genehmigung der University of Pittsburgh. Alle Rechte vorbehalten,

25. HR-013-24-04 F. Meiner an H. Reichenbach am 2, 6, 1937. Zitiert mit Genehmigung der University of Pittsburgh. Alle Rechte vorbehalten.

wurde die " Erkenntnis " schließlich Ende des Jahres 1937 an Van Stockum & Zoon in Holland verkauft und zugleich von The University of Chicago Press vertrieben. Für den Band 7 (1937/38) zeichnete Carnap als alleiniger Herausgeber.

Der einstmals so engagierte Reichenbach selbst meldete sich im Dezember 1937 aus der Emigration in Istanbul im Hinblick auf die weitere Entwicklung der Erkenntnis noch einmal zu Wort. Er schrieb ganz im Sinne seiner bisherigen Auffassungen an Charles Morris, Professor für Philosophie an der Universität Chicago, der sich in jener Zeit sehr für die Berufung Reichenbachs (dessen Vertrag mit der türkischen Regierung lief 1938 aus) nach Los Angeles einsetzte : *I heard from Meiner that you proposed to use as a subtitle the term " Unity of Science ", with the intention to use it later alone. I am decidedly against this title or subtitle. The word Unity of Science does not at all express what we want. ... The title goes back to the old Vienna idea, derived from Wittgenstein, that there is no science of philosophy, or no special subject which can be discussed by us. ... The unity of science is not our program, but a special thesis maintained by some among us, or even by all of us if the term is sufficiently widely interpreted. As a program it would mean : calling all men of science together to cooperate for their special purpose, ...*

If we invite men of science to cooperate this is always in the special purpose to discuss the foundations of knowledge. Thus what characterises our program is the study of the foundation of knowledge...

If we use an English title I suggest " Journal of Logistic Empiricism ", or also " (...) of scientific empiricism. ... "[26].

Anfang 1938 schrieb Reichenbach an Neurath — informierende Briefe waren wohl verloren gegangen — " Schade, daß Morris in der Erkenntnissache keine Fortschritte gemacht hat. Es wird wohl nun dazu kommen, daß die " Erkenntnis " ganz eingeht "[27]. Doch noch erschien Band 8 (1939-1940) unter dem Titel *The Journal of Unified Science* (*Erkenntnis*), wieder mit Reichenbach und Carnap als *Editors* und sechs *Associate Editors*. Nach der Besetzung der Niederlande durch deutsche Truppen war das Erscheinen weiterer Bände allerdings nicht mehr möglich, und auch dem Wunsch Reichenbachs, in den USA eine neue Zeitschrift zu gründen, war kein Erfolg beschieden.

Erst Jahrzehnte später sollte es doch eine Weiterführung der Erkenntnis geben ! Nach Hegselmann und Siegwart wurden bereits Mitte der sechziger Jahre " im Kreis um Wolfgang Stegmüller in München verschiedene Konzepte zur Gründung einer Zeitschrift für analytische Philosophie erwogen. Es sollte ein Forum analytischen Denkens geschaffen werden, das einerseits einem drin-

---

26. HR-013-50-47 H. Reichenbach an C. Morris am 1, 12, 1937. Zitiert mit Genehmigung der University of Pittsburgh. Alle Rechte vorbehalten.

27. HR-013-51-19 H. Reichenbach an O. Neurath am 7, 2, 1938. Zitiert mit Genehmigung der University of Pittsburgh. Alle Rechte vorbehalten.

genden Desiderat im europäisch-kontinentalen Raum Abhilfe schaffen, andererseits gleichwohl aber einen internationalen Einzugsbereich besitzen sollte. Zur Realisierung gelangte schließlich die Idee einer Fortsetzung der Erkenntnis, die ... auch vom Verlag Meiner seit längerem in Betracht gezogen worden war. ... Carl G. Hempel, Wolfgang Stegmüller und Wilhelm K. Essler brachten 1975 mit dem 9. Band die neue *Erkenntnis* auf die Bahn. ...

In seiner programmatischen Einleitung stellte Hempel ... unmißverständlich klar, daß die philosophische Entwicklung seit der Einstellung der Erkenntnis es ausschließt, daß es der Nachfolgezeitschrift um die Kultivierung der spezifischen Doktrin des Logischen Empirismus zu tun ist : die Fortschreibung dieser Strömung selbst habe dazu geführt, daß viele Ideen aufgegeben oder modifiziert werden mußten und neue hinzugekommen sind. ... Was indes Reichenbach den Autoren der alten Erkenntnis abforderte, machte Hempel auch für die Autoren der neuen verpflichtend : *adherence to high standards of clarity of statement and cogency of reasoning* ˮ[28].

---

28. R. Hegselmann, G. Siegwart, " Zur Geschichte der 'Erkenntnis' ˮ, *Erkenntnis : An International Journal of Analytic Philosophy,* vol. 35, 1-3 (1991), 470.

KUHN'S PHILOSOPHICAL TROUBLES WITH ACTUAL

HISTORY OF SCIENCE[1]

Mario H. OTERO

Almost everybody knows that Thomas S. Kuhn was both an historian of science and a philosopher of science. According to him he is not, nor was, both at once. It seems that it is possible for him to distinguish when he is either one. Even more, he thinks that both enterprises should be separated[2] not only in his work but also in general notwithstanding the mutual fertilization between them.

Many of us estimate very doubtful that the separation be present even in Kuhn's intellectual practice[3], in SSR[4], his main and most famous work, we do not find such alleged distances.

Concerning this question we would like to compare here some passages of his recent paper on " The trouble with the historical philosophy of science "[5] with others of the former one on " The relations between the history and the philosophy of science "[6], included in the book *The essential tension*[7].

1. See M.H. Otero, " Apuntes sobre el último Kuhn ", *Llull*, vol. 19 (1996).

2. R. Stuewer (ed.), *Historical and philosophical perspectives of science*, New York ; First edition in *Minnnesota Studies in the philosophy of science*, vol. 5, Minneapolis, 1970, deal extensively with the subject of the " distance " or " divorce " between history and philosophy of science.

3. F. Zamora, " El último Kuhn ", *Arbor*, vol. 148 (1994), discusses important aspects of historico-philosophical practice in Kuhn's last period though not specifically about his theory on the relations between them.

4. T.S. Kuhn, *The structure of scientific revolutions*, 2nd edition, Chicago, 1970.

5. T.S. Kuhn, *The trouble with the historical philosophy of science*, Cambridge, MA, 1992, T from now on.

6. M. Wartofsky, " The relation between philosophy of science and history of science ", in R.S. Cohen, P.K. Feyerabend, M.W. Wartofsky (eds), *Essays in memory of Imre Lakatos*, Dordrecht, 1976 (1st ed. 1968), from now on, R.

7. T.S. Kuhn, " The relations between the history and the philosophy of science ", *The essential tension*, Chicago, 1977 (conference delivered in 1975).

We should remember the double autobiographical character of R, in aspects concerning Kuhn's own formation and activities and his long experience in the teaching of both disciplines and in the orientation of doctoral theses.

Even more, the last several Kuhn papers of the nineties[8] have also a strong autobiographical character.

It would seem that the original position in R — not totally exempt of ambiguities — could have been broken by his practice and that in T there would appear some very surprising theses.

1. As for T, philosophical construction seemed to be attained, in Kuhn's original generation, from observations of scientific actual behaviors. But for him that image is misleading because in that historical philosophy of science conclusions may be reached with scarce reference to real historical records. Even more, the historical perspective, following T, was in the beginning alien to the received and dominant philosophical tradition that was guided rather by the existence, or not, of a rational guarantee as a basis to affirm this or that. For Kuhn gradually the static image of the tradition became to be dynamic in the new philosophy and science began to be conceived as a developmental practice or enterprise. Even the attained new perspective could be derived *from principles* and not necessarily from historical records[9].

" Now I think we overemphasized the empirical aspect of our enterprise " (T, 6).

and so, because the point of departure were principles, one may explain for Kuhn the scarce contingence of consequences, " ...making them harder to dismiss as a product of muckraking investigation by those hostile to science " (T, 10).

2.1 The result of the historian activity would be *a narrative* that would include a description of the initial state of the process to be explained. It would include also a description of the beliefs at that moment and of the conceptual vocabulary in use. Those resulting considerable changes at the end of the process would come from intermediate and not too notorious gradual changes. What goes on in the process would be a change of beliefs within changes in the context. Concerning the former ones it would be necessary to investigate precisely why the actors decided those changes.

---

8. T.S. Kuhn, " The Road since *Structure* ", in A. Fine, M. Forbes, L. Wessels (eds), *Proceedings of the 1990 Biennial Meeting of the Philosophy of Science Association*, vol. 2 (1991) ; T.S. Kuhn, *The trouble with the historical philosophy of science*, Cambridge, MA, 1992 ; T.S. Kuhn, " Afterwords ", P. Horwich (ed.), *World changes ; Thomas Kuhn and the nature of science*, Cambridge, MA, 1993 ; T.S. Kuhn, " History of science ", in P.D. Asquith, H.E. Kyburg (eds), *Current research in philosophy of science*, East Lansing, MI, 1979.

9. Nevertheless the reciprocal influence between history and philosophy of science is clear not only in *The structure of scientific revolutions*, but also in *The copernican revolution*. Still more, many other Kuhn books, papers, reviews and short notes on historical subjects, listed in P. Hoyningen-Huene, *Thomas S. Kuhn's philosophy of science*, Chicago, 1989, are not alien to the theme of the referred reciprocal influence.

2.2 For the philosophers[10] the problem would be the same : that is, to understand small changes in beliefs. Rationality, objectivity and evidence would come to be subjects easier to deal with that with the referents of the corresponding beliefs. The static Archimedian platform required by the so called neutral observation in the former tradition was then unnecessary and it would have vanished.

First of all, as for Kuhn, the rationality in historical perspective needs a transitory rationality only in relation with the members of the group which produces each decision. Secondly the changes to evaluate are always relatively small even if they may seem gigantic in retrospect. Thirdly, in general truth would not come from of comparing beliefs with reality : the evaluation would be indirect. The criteria that intervene are secondary criteria : precision (only approximate and often unattainable), consistence with other accepted beliefs (at most local), breadth of applicability (increasingly narrow when time goes on), simplicity (depending on the observing eye), among others. They are ambiguous values that anyway are not satisfied at once. But if those criteria were applied to belief changes they would get, for Kuhn, new relevance and sense, both relational ones : a set of beliefs may become more precise, more consistent, larger in applicability, more simple, *without becoming truer* (T, 13-14).

The expression 'truer' is sometimes interpreted as 'more probable' but that would carry, even in this Kuhn, what has received the name of 'disastrous metainduction' (as Kitcher baptised it) : " All past beliefs about nature have sooner or later turned out to be false... the probability that any currently proposed belief will far better must be close to zero " (T, 14).

Chilling result, and erroneous from my point of view ; already discarded by Poincaré, not without good reasons, at the beginnings of the century.

The disastrous metainduction would complement in this way, even radicalizing it, the so recurred underdetermination of theory.

The consequences that Kuhn presents have even a larger scope : " I am not suggesting, let me emphasize, that there is a reality which science fails to get at. My point is rather that no sense can be made of the notion of reality as it has ordinarily functioned in philosophy of science " (*ibidem*).

Amazing ...

10. Not only " The trouble with the historical philosophy of science " raises the subject of the philosophical enterprise of those occupied with science ; also " Dubbing and redubbing... " and T.S. Kuhn, " Possible worlds in history of science ", in S. Allén (ed.), *Possible worlds in humanities, arts and sciences*, Berlin, 1989, raise it, in a somewhat but not essentially different version of the former. In both Kuhn elaborates on the natural class concept and on local holism. T.S. Kuhn, " The natural and the human sciences ", in D.H. Hiley, J.E. Bohman, R. Shusterman (eds), *The interpretive turn*, Ithaca, 1991 and 1993 — this written earlier than T —, also work on the subject of that philosophical enterprise.

Kuhn, as he says, is not far of the strong programme[11] : " ...facts are not prior to conclusions drawn from them and those conclusions cannot claim truth " (ibidem).

A final confession, advanced earlier as a sketch, is especially clarifying : " I've reached that position from principles that must govern all developmental processes, without, that is, needing to call upon actual examples of scientific behavior " (ibidem).

Sensational, then history of real science, what for ?

Towards the end of T. Kuhn returns to its central subject.

The trouble with the historical philosophy of science comes for him from the fact that its quasihistorical or perihistorical examples have questioned the authority of science itself. The pillars of that authority. 1. The priority of facts and its independence from the consequences and 2. The truths concerning an independent external world — would have melt. The option Kuhn faced was either to provide them a firm foundation or to eliminate them completely. But now he maintains that what matters are not observed facts concerning scientific practice but necessary characteristics owned by the *evolutionary* processes in general. Should we think that in such way Kuhn's difficulty — a quite persistent and enough annoying one — would be totally overcome ?

3. From the early R — very rich and at the same time questionable text — we will take only one point, leaving for some other opportunity other very interesting aspects.

When Kuhn strongly doubts about the value of the covering law model for history (R, 15-16), his central criticism points to the triviality in some cases, or the non historical character in others (sociological aspects or belonging to social sciences), of the laws that would be assumed by the historian in that model[12]. To suppose those laws would amount to force the historian to employ instruments totally alien and of doubtful validity for accomplishing his job.

Then we could demand ourselves if the principles and examples quasi- or perihistorical that Kuhn prefers for the historical philosopher of science would not be purely speculative, because, avowedly, they renounce both to empirical

11. See M.H. Otero, " Apuntes sobre el último Kuhn ", *op. cit.*, 1996 and C. Solís, *Razones e intereses : la historia de la ciencia después de Kuhn*, Barcelona, 1994.

12. It is enough evident that Kuhn alludes to the well known Hempel paper " The function of general laws in history ", *The Journal of Philosophy*, vol. 39 (1942). Shortly later Theodor Abel, presented a very intelligent contribution in " The operation called Verstehen ", *American Journal of Sociology*, vol. 54 (1948). After a lapse of large domination of the covering law model, with its well known sequels, appeared often the criticisms that, in many cases, arrived to a notion very close to that of Verstehen, the very notion that Hempel had tried to supersede. Von Wright presented in his " Explanation and understanding " a new paradigmatical concept. But he didn't go back to the diltheyian and marburgian Verstehen. Kuhn was strongly influenced by this new orientation. Each time Kuhn used the renewals produced in the hardware of the orthodox-analytic philosophy and then he produced the corresponding rectifications in his thought.

test and to actual historical records and explanations (we must remember that for Kuhn historical work needs not to be only descriptive).

4. Even if we have considered here only limited aspects of R and T, consistent with many other not alluded passages of those texts and of others, we may point the origin of our strong surprise concerning the central thesis included in T.

a. For Kuhn history of science and philosophy of science are different things even if they fertilize each other,

b. Kuhn's practice in his main works, and especially in *SSR*, seems to be different to the conception exposed in R (and obviously in T), with a strong overlapping if not integration of both supposed separate disciplines,

c. The independence — so it seems in the texts — of the historical theses belonging to philosophy of science (hypostatiated principles and examples) and opposed to the results of actual history of science, far from immunizing those theses extremely weakens them, and

d. Kuhn would not be situated in such way, from the comparison of his own words, far neither from the " deconstruction gone mad " of the strong program of the sociophilosophy of knowledge nor from the constructivist-idealist[13] theses that Edouard LeRoy exposed almost a hundred years ago.

13. Constructivist and even idealist modes appear in the niche idea at the end of T ; see P. Hoyningen Huene, *Thomas S. Kuhn's philosophy of science*, *op. cit.*, and M.H. Otero, " Apuntes sobre el último Kuhn ", *op. cit.* ; M.H. Otero, " Tres modalidades de inmanentismo ", *Diánoia* ; D. Peral, P. Estévez, A. Pulgarín, " Presencia del pensamiento kuhniano en la literatura científica : 1966-1995 ", *Llull*, vol. 20 (1997).

ESSAI SUR LA NATURE DES THÉORIES SCIENTIFIQUES[1].
THÉORIES ET STRUCTURES

Jean GADISSEUR

L'homme, être connaissant, est fasciné par le miracle de sa propre connaissance. Cherchant à la comprendre, et en faisant donc un objet de connaissance, il s'aventure sur le sentier abrupt de l'épistémologie, manifestement la science la plus difficile qui soit.

Ce n'est pas pourtant qu'elle devrait l'être, puisque son objet, par sa nature même est parfaitement connu. Il n'y a pas, en matière d'épistémologie, de véritable secret à découvrir ; il n'y a pas de fluide ignoré ni de force inconnue qu'il

1. J'espère qu'on me pardonnera d'avoir, en écrivant ce texte, sciemment omis presque toute référence. C'est qu'il les aurait fallu trop nombreuses : on sait que le sujet n'est pas neuf. S'engager dans la voie des comparaisons et des arguments magistraux aurait conduit trop loin. Quant au mérite des bonnes idées que cet article peut comporter, si l'on jugeait qu'il y en a et si l'on estimait qu'elles sont porteuses de mérite, je demanderais qu'on l'inscrive au crédit de l'Humanité. Il y a trop de César à qui rendre leur dû. On pourra évidemment m'attribuer, et à moi seul, la paternité des erreurs qui ne manquent sans doute pas — il importe peu. Je ne me sens ici qu'un seul vrai devoir de justice, vis-à-vis d'un homme qui a été et est encore mon maître. Lorsque s'est constitué en 1967 sous l'impulsion et la direction de Pierre Lebrun, le Groupe d'histoire quantitative de l'Université de Liège, nous étions loin de soupçonner que la difficulté majeure à laquelle nous allions nous heurter était de nature épistémologique. Nous étions confiants que les sciences objectives de l'homme — économie, politologie, démographie, sociologie — allaient nous procurer les méthodes et les instruments qui nous permettraient d'éclairer l'histoire d'une lumière singulièrement nouvelle. Nous pensions qu'il suffirait de puiser dans l'arsenal méthodologique et théorique qu'elles nous proposaient afin d'apporter au réel historique une véritable explication scientifique, qui transcendât enfin celle, traditionnelle et combien décevante, des enchaînements de faits particuliers. Il nous a fallu déchanter, et notamment parce que la théorie économique n'est pas transposable à la réalité historique, ni pour en décrire le fonctionnement, ni — encore moins — pour rendre compte de son changement. Lorsqu'il fallut expliquer la Révolution Industrielle, Pierre Lebrun eut l'idée de représenter le réel historique en termes de structures, et de rendre compte de leur changement par la structure génétique, soit une structure de changement de structure. Il marquait ainsi le point de départ d'une réflexion qui devait nous conduire à reconsidérer, en une démarche commune, mais lente et difficile, non seulement l'histoire ou la science économique mais encore les conceptions épistémologiques les plus couramment admises. Les quelques idées que je présente ici ne sont que des conséquences de cette véritable invention. Quant aux résultats complets de la démarche épistémologique, on les trouvera en leur temps dans *D'une histoire l'autre*, tome I de la collection *Histoire quantitative et développement de la Belgique*, éditée par l'Académie Royale de Belgique.

s'agirait de mettre au jour, ou d'organisation sibylline qu'il faudrait élucider par une observation minutieuse et patiente du réel, par l'une ou l'autre intuition géniale. Il y a au contraire des représentations, et notamment des théories scientifiques qui, artéfacts humains, n'ont évidemment rien de mystérieux pour l'homme, à ceci près que, les construisant presque comme il respire, il voit bien qu'il les élabore, mais ne comprend que malaisément ni comment il les produit, ni en quoi elles consistent.

Cette difficulté paradoxale résulte sans doute d'une multitude de causes. Un des principaux obstacles est que l'épistémologie relève de la conscience ; elle est donc une science subjective, et donc soumise à une foule de préjugés inconscients qui forment l'opinion que l'homme a de lui-même — notamment du miroir de sa propre conscience. Ainsi ne faut-il pas sous-estimer l'inévitable confusion qu'il établit, dans *sa praxis,* entre le réel et les représentations qu'il en forme, ni la fiction de rationalité, à laquelle il ne peut guère rationnellement renoncer, ni l'inévitable anthropomorphisme de sa pensée et du langage qui la supporte et l'exprime, ni encore la tendance naturelle de son esprit à vouloir simplifier et généraliser trop rapidement les choses, ce qui conduit à un dommageable réductionnisme.

Une autre difficulté grave résulte de la dimension historique de la démarche épistémologique, et par conséquent de sa focalisation sur les constructions des sciences qui se sont développées le plus précocement et le plus spectaculairement — il m'est arrivé à ce propos de parler durement, quitte à choquer quelque peu, de l'impérialisme épistémologique du paradigme physico-mathématico-expérimental. Mais il est bien vrai que les théories de la physique ont inspiré de manière prédominante toute la réflexion épistémologique, et continuent de le faire. En particulier, la mécanique ou la thermodynamique se sont imposées comme modèles de simplicité, d'élégance et d'efficacité. Leurs constructions sont admises, implicitement ou explicitement, comme les formes idéales auxquelles toutes les sciences du réel doivent tendre — même celles de l'homme. Qu'il me suffise de rappeler le voeu d'un A. Comte ou d'un A. Quételet pour la construction d'une " physique sociale ".

Or, considérant ainsi la science[2], j'affirme qu'on n'en a vu qu'une petite partie, un seul aspect. Le paradigme épistémologique physico-mathématique ne correspond en effet qu'aux théories simples de la physique, celles qui rendent compte des propriétés de la matière. Il s'agit de *théories scientifiques restreintes* qui consistent en la réduction axiomatique des lois décrivant un seul type de comportement d'une seule catégorie d'objets. Par opposition, la cons-

---

2. Je ne parlerai ici que des sciences qui s'attachent à décrire le réel, soient celles qui étudient la nature et celles qui étudient l'homme en tant que réalité objective. Quant aux mathématiques et à la philosophie — dont l'épistémologie —, qui étudient les construits de l'esprit ou de la conscience, il s'agit bien sûr d'autre chose. Il pourra encore y être question de théories et de structures, mais de théories et de structures d'une nature changée par celle de l'objet auquel elles s'appliquent.

truction normale de toute connaissance scientifique, celle qui appréhende le réel dans sa complexité, est la *structure scientifique,* ou la *théorie scientifique au sens large.*

L'erreur a été de considérer toutes les sciences du point de vue de la physique, alors qu'il aurait fallu faire l'inverse, et considérer les théories restreintes de la physique comme des cas particuliers de constructions scientifiques : ceux pour lesquels la structure scientifique s'anamorphose par simplification extrême en théorie scientifique restreinte. L'erreur n'était que de perspective, mais elle a produit des conséquences graves.

Alors que l'épistémologie devait fournir aux sciences — à toutes les sciences — des schèmes fondamentaux, des modèles normatifs capables d'orienter efficacement leur démarche, elle leur a proposé un type théorique largement inadéquat. On peut croire que les sciences du vivant n'en ont que relativement peu souffert, quoique pendant longtemps la volonté de croire en un principe vitaliste, évidemment inspiré des principes ou axiomes de la physique, les ait certainement empêchées de progresser[3]. Le véritable drame a été celui des sciences humaines. Elles n'ont pu que tenter de se plier au paradigme épistémologique, comme l'a fait l'économie — mais ce paradigme inadéquat n'a conduit qu'à des constructions insuffisantes —, ou le rejeter en affirmant une autre spécificité scientifique, comme l'a fait l'histoire — mais alors, elles se privaient de la simplification et de la généralisation propres à la science. D'une façon comme de l'autre, les sciences humaines ont végété. Or, s'il est un domaine de connaissance où il est désespérément urgent de progresser, c'est bien celui de l'homme lui-même. Il a réalisé dans les sciences de la matière, d'abord inerte et maintenant vivante, des progrès tellement rapides qu'il n'arrive plus, et depuis trop longtemps déjà, à en maîtriser les usages. Devenu capable de soumettre à sa volonté son environnement matériel, l'homme en vient, par ignorance de ce qu'il est et de ce que sont ses sociétés, à se changer lui-même sans le vouloir, et sans prévoir comment ni jusqu'où.

On jugera peut-être que l'ambition des pages qui suivent est fort grande. Il est vrai que les quelques idées, au demeurant simples, que je vais tenter d'y exposer peuvent conduire à réunifier le champ scientifique, ou être les fondements d'une théorie épistémologique capable peut-être d'intégrer de manière cohérente les acquis scientifiques réalisés jusqu'à présent comme de proposer des modèles épistémologiques renouvelés. Dans le bref espace qui m'est ouvert, je ne pourrai cependant pousser bien loin l'explication ni développer beaucoup les conséquences de ce que je me propose de montrer. Le lecteur attentif, mais indulgent, me pardonnera sans doute la brièveté et l'aridité du

---

3. Le génie de Claude Bernard a sans doute été de composer avec ce paradigme physico-mathématique en proposant une méthode qui attaque le problème par l'autre bout : celui de l'observation du réel, plutôt que l'établissement de principes réducteurs. Le prestige de la médecine, et donc des sciences du vivant a contribué à intégrer la règle de l'expérimentation au paradigme épistémologique, pour en faire le paradigme physico-mathématico-expérimental.

propos. J'espère seulement qu'ayant entrebâillé pour lui une porte, il apercevra ou devinera la route sur laquelle je la crois s'ouvrir.

LE POINT DE VUE MICROSCOPIQUE ET LE POINT DE VUE MACROSCOPIQUE

Préalablement à toute investigation scientifique, il y a un découpage du réel. Il faut y distinguer des objets ou des catégories d'objets, et le faire évidemment de façon adéquate. Le découpage est spatial et temporel ; il doit conduire à délimiter des objets homogènes et stables. Leur nature ou leur dimension peut être quelconque. Une fois le découpage effectué, l'objet peut être envisagé de deux manières, alternativement, selon deux points de vue.

Le premier de ces points de vue est le point de vue microscopique : on regardera l'objet comme un tout insécable, une sorte de boite noire — à proprement parler un *atome* — dont il s'agira seulement d'observer et d'expliquer les comportements externes, soient les relations qu'il entretient avec son environnement — y compris l'observateur lui-même. Qu'il s'agisse par exemple d'une plante, il faudra considérer sa germination, sa croissance, sa floraison, les flux qu'elle entretient avec son milieu, ses changements d'apparence ou de forme selon le rythme des jours ou des saisons. Ou bien, s'il s'agit d'un corps au sens de la mécanique, c'est-à-dire un objet massif localisé dans l'espace et doté d'un mouvement, on observera son mouvement relativement aux autres corps qui l'environnent.

Je caractérise ce point de vue de " microscopique "[4], car l'objet considéré est petit par rapport à un environnement plus vaste avec lequel il interagit.

A ce stade, il n'est donc pas question de mettre en relation les comportements externes avec un quelconque fonctionnement interne dont il serait le résultat. On se borne à étudier les comportements externes de l'objet, que l'on traduit en lois scientifiques, c'est-à-dire en propositions qui en décrivent les régularités.

4. Selon cet emploi des termes, le microscope des biologistes devrait s'appeler " macroscope ", car il permet de voir grand — en photographie, on utilise bien des objectifs " macro " afin de fixer l'image des petites choses ; dans le même esprit, le télescope est mal nommé, car il n'éloigne pas, il rapproche. Cette remarque afin d'éviter la confusion des idées : l'approche microscopique, c'est regarder par le gros bout de la lunette, afin d'élargir le champ de vision. Regarder au travers d'un microscope, c'est adopter un point de vue macroscopique. Il me faut insister sur la relativité de la distinction que j'introduis : l'astronome qui observe le comportement externe d'Antarès se place d'un point de vue microscopique. Le microbiologiste qui étudie la structure d'un virus l'observe macroscopiquement, fût-ce grâce à un microscope électronique. Une remarque du même genre peut être faite à propos de l'économie. Il y a d'une part la micro-économie, qui étudie le comportement des agents — firmes, individus ou ménages —, il y a d'autre part la macro-économie qui étude le fonctionnement des sociétés économiques. Dans ce dernier cas, le point de vue adopté se veut effectivement macroscopique, car c'est le fonctionnement de la société économique qui est visé. Dans le premier cependant, le point de vue sera tantôt microscopique — c'est le cas de l'analyse du comportement de l'agent individuel dans l'environnement que constitue l'économie de marché —, tantôt macroscopique lorsqu'il s'agit d'étudier l'organisation et le fonctionnement interne des entreprises.

Quant au point de vue macroscopique, il est évidemment complémentaire au premier. Ici, l'objet occupe tout le champ de vision de l'observateur, car c'est maintenant son organisation, son fonctionnement interne qui est étudié. L'objet, quelles que soient d'ailleurs ses dimensions absolues, occupe toute le champ d'observation d'où le terme de " macroscopique ". Dans le cas de la plante, on observera les comportements et les interactions des éléments qui la composent. Il s'agira d'organes : racines, tronc, tiges, feuilles, bourgeons, etc. ; ou de cellules spécialisées, ou encore des éléments qui constituent les cellules et participent à leur fonctionnement.

On voit bien que les approches micro- et macroscopique comportent des étages des niveaux d'analyse qui peuvent être fort différents, mais qui cependant ne sont pas indépendants et correspondent à la dimension du découpage des objets étudiés. On peut ainsi considérer la plante comme un ensemble d'organes, dont les comportements externes et les interactions expliquent le fonctionnement de l'objet complexe qu'elle constitue. Mais les organes peuvent à leur tour être considérés comme des objets complexes dont il s'agit encore d'étudier le fonctionnement. On les analysera alors comme l'organisation de cellules de différents types, chacune dotée des comportements caractéristiques de la classe à laquelle elle appartient.

Puis les cellules elles-mêmes pourront être envisagées comme des objets complexes dont il faudra étudier l'organisation, et donc dans lesquelles on distinguera à nouveau des éléments constitutifs, évidemment plus petits, comme le noyau, le cytoplasme, la membrane, les mitochondries, les vacuoles et autres lysosomes.

Mais nous pouvons aussi nous déplacer dans l'autre direction et, plutôt que de décomposer la plante en ses éléments, la voir comme un élément de l'ensemble plus vaste que constitue l'écosystème. Cet écosystème lui-même pourra être envisagé comme l'élément d'une évolution géologique, si nous choisissons d'étudier un ensemble qui soit plus grand, non suivant les dimensions ordinaires de l'espace, mais suivant cette fois celle du temps historique.

Ainsi, l'investigation scientifique peut être comprise comme une succession d'approches alternativement microscopiques et macroscopiques. L'approche macroscopique expliquant le comportement externe — donc microscopiquement constaté — de l'objet envisagé dans sa complexité à partir des comportements externes des éléments qui le composent. On peut voir au processus deux limites, celle du plus grand, soit l'Univers, pour lequel le point de vue microscopique n'a pas de sens, faute d'un environnement dans lequel le situer, et l'objet ou la particule les plus élémentaires, c'est-à-dire dénués de toute complexité, et donc non observables du point de vue macroscopique. Entre ces deux extrêmes, une infinité de niveaux d'analyse ou d'observation sont possibles, en fonction de l'arbitraire des découpages.

Selon l'approche adoptée, la construction théorique sera évidemment différente. Le point de vue microscopique pourra conduire à l'élaboration de *théories scientifiques restreintes.* L'approche macroscopique devra conduire à la construction de *structures scientifiques,* ou *théories scientifiques au sens large.*

LA THÉORIE SCIENTIFIQUE RESTREINTE

Dans certains cas — mais relativement rares — l'approche microscopique peut conduire à une réduction axiomatique des lois observées de manière à former une théorie scientifique restreinte. Le paradigme en est la théorie de la gravitation universelle de Newton, ou la théorie de la relativité, qu'elle soit ou non généralisée.

La théorie scientifique restreinte ne concerne qu'une seule catégorie d'objets et, pour ces objets, un seul type de comportement. C'est-à-dire qu'au découpage du réel en objets s'en superpose un autre qui conduit à la sélection d'une seule catégorie de comportements — découpage méthodologique ou disciplinaire.

On peut énoncer, suivant le schéma de la fig. 1, la démarche logique en quatre étapes, hormis les découpages initiaux que nous considérerons comme acquis.

1°   L'observation des objets suivant un type de comportement conduit à l'énonciation d'un ensemble de lois scientifiques correspondant à autant de régularités. Ces lois décrivent non seulement le comportement de l'objet, mais évidemment aussi les circonstances externes qui l'accompagnent — je dirai que les lois scientifiques sont corrélatives. Elles ont un plus ou moins grand degré de généralisation et donc d'abstraction. Néanmoins, elles peuvent paraître n'avoir entre elles que peu de parenté, et se présenter donc comme un ensemble disparate. Les lois de la balistique, celle de la chute des corps, les lois qui décrivent le mouvement des corps célestes, celles qui règlent le mouvement des marées, la relation entre l'altitude et la pression atmosphérique, le principe d'Archimède, etc., semblent effectivement autant de comportements distincts des corps physiques, au sens de la mécanique.

2°   Une intuition aussi géniale que simplificatrice conduit à les considérer comme l'expression particulière d'un nombre limité de principes très généraux. Dans le cas de Newton, il s'agit du principe de la gravitation universelle et de celui de l'inertie, plus un ensemble d'hypothèses destinées à assurer la maniabilité du système, mais que l'on peut considérer comme approximativement vraies.

Ces principes sont abstraits, et ne sont donc pas directement observables[5]. On ne peut en percevoir les effets qu'au travers des lois, donc des comportements qui paraissent en être les conséquences. Puisque la science n'est pas une religion, quoique nous pensions et quoique nous nous comportions souvent comme si c'était le cas, il n'y a évidemment pas lieu de croire qu'il existe quelque chose comme une force gravitationnelle, ou une véritable inertie. Invoquer l'une et l'autre est tout simplement commode et efficace.

Ces principes, en réalité sont seulement des axiomes, soit des propositions qui constituent le point de départ, eux-mêmes non démontrables déductivement, d'un ensemble de déductions parmi lesquelles figurent, à un degré de correction près, les lois empiriquement observées.

3° La théorie scientifique restreinte est donc constituée des axiomes qui la fondent et de l'ensemble des déductions (lois théoriques) auxquelles ils donnent lieu. Le miracle scientifique, et ce qui a considérablement contribué au prestige des théories scientifiques restreintes élaborées notamment en mécanique, c'est que la construction est féconde. En effet, elle permet deux choses, en tous cas à l'intérieur de certaines limites, qui sont celles de sa validité : (a) reformuler en les précisant et éventuellement en les corrigeant les lois empiriques de départ ; (b) déduire des comportements qui n'avaient pas encore été constatés, soit parce qu'ils se situent au-delà des limites du directement observable, soit par défaut d'un système conceptuel adéquat. La fécondité de la théorie scientifique restreinte consiste donc en ce qu'elle permet de substituer aux lois empiriques de départ, incomplètes, approximatives, formulées de façon hétérogène, un ensemble de lois théoriques plus nombreuses, plus précises — exactes dira-t-on parfois, mais le terme est abusif — et homogènes.

4° Enfin, ces lois rénovées ou nouvelles seront confrontées au réel. Il ne s'agira pas ici de méthode expérimentale ; il ne s'agira pas de vérifier ou d'infirmer la théorie — n'est-elle pas évidemment vraie relativement aux lois dont elle rend compte ? — mais bien d'en mesurer le domaine de validité. Ainsi la théorie de Newton n'a jamais été et ne pourra jamais être infirmée ou confirmée par l'observation ou l'expé-

5. J'ajouterai qu'ils sont absolument dépourvus de réalité. Il n'existe pas quelque chose qui soit la gravitation, ou l'inertie. Il n'existe même pas quelque chose qui soit une force d'attraction. Les termes eux-mêmes ne traduisent que l'anthropomorphisme de nos représentations. Je sais comme chacun, par expérience personnelle, que nous avons naturellement tendance à croire en l'existence des forces, et il est vrai qu'elles ont pour l'homme un degré d'existentialité. Mais on peut bien sûr représenter autrement les phénomènes. Qu'on imagine par exemple que le champ gravitationnel, au lieu d'exercer une force sur les corps, infléchit la fonction de probabilité qui régit le mouvement brownien des particules qui les composent ; ou bien encore que le phénomène résulte de l'interaction des fonctions d'onde qui représentent les particules. La conséquence est évidemment la même : les représentations sont différentes. Cependant, aucune n'est plus vraie qu'une autre. Nous sommes bien sûr toujours dans le domaine du " comme si ".

rience. Simplement a-t-on été amené à constater que son domaine de validité se restreignait aux corps dotés de vitesses lentes, et à la compléter en conséquence du principe de relativité afin d'en étendre l'application aux vitesses quasi-luminiques.

FIGURE 1. THÉORIE SCIENTIFIQUE RESTREINTE.

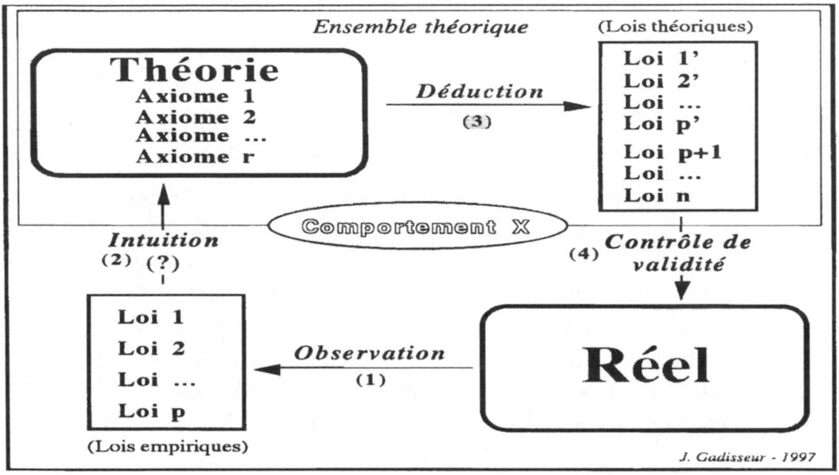

La théorie scientifique restreinte est l'aboutissement de l'approche microscopique de la réalité, qui envisage seulement les comportements externes des objets situés dans un environnement avec lequel ils interagissent. La théorie scientifique restreinte ne concerne qu'une seule catégorie de comportements appartenant à une seule classe d'objets. Elle comporte une réduction axiomatique — théorie proprement dite — des lois empiriques qui en forment le point de départ. Déductivement elle permet de reformuler et de corriger ces lois, et de prévoir de nouveaux comportements non encore observés. Cependant, sa validité est toujours limitée.

Cette schématisation de la construction des théories scientifiques restreintes appelle de nombreuses remarques. Je me contenterai d'en évoquer ici trois.

En premier lieu, le schéma proposé est purement abstrait, et ne correspond évidemment pas à la réalité de la démarche. Le caractère essentiellement dialectique de tout processus de connaissance fait qu'il faut impérativement disposer d'un cadre théorique afin d'observer le réel. Ce que j'ai appelé " lois empiriques " comporte donc déjà nécessairement une part de théorie. On pourrait même dire que la théorie restreinte est déjà implicite, quoique approximative, dans toute observation du réel, et que la démarche ne consiste en fait qu'à l'expliciter, à la formaliser et à la compléter.

En second lieu, une théorie scientifique restreinte peut parfois être prise en compte par une autre théorie scientifique restreinte plus générale, capable d'unifier une plus grande variété de comportements. Ainsi par exemple la mécanique ondulatoire englobe-t-elle dans son application déductive la théorie de l'optique. De la même manière que des lois empiriques peuvent être réduites à quelques principes généraux par l'intégration théorique, des comportements qui paraissent différents peuvent être ré-duits à des manifestations d'un comportement plus fondamental. Il s'agit bien de réductionnisme, mais d'un réductionnisme légitime, car les constructions auxquelles il conduit ne perdent rien de la description opérée par les lois qu'elles intègrent. Il y a au contraire, sans compter l'avantage d'une plus grande élégance des représentations, un double gain : celui d'une plus grande simplicité et celui apporté par la fécondité de la démarche, que l'on peut espérer d'autant plus grande que l'intégration est plus vaste.

En troisième lieu, comme je l'ai déjà évoqué, la construction de théories scientifiques restreintes ne peut concerner qu'une seule catégorie de comportements, ni donc qu'une seule classe d'objets. Ce sont les corps de la mécanique, considérés dans leur masse et leur mouvement, ou ceux de la thermodynamique considérés dans leur énergie sous ses différentes formes ; ce sont les atomes de Mendeleïev du point de vue de leur valence. L'approche est bien entendu généralisante et féconde, mais n'est utilement applicable qu'aux comportements les plus élémentaires du réel. Vouloir notamment construire à propos des comportements du vivant ou de l'humain de telles théories restreintes ne peut conduire qu'à l'échec, puisqu'elles ne pourront rendre compte de la complexité essentielle de l'objet, soit de son organisation[6].

LA STRUCTURE SCIENTIFIQUE

On ne peut donc évidemment se borner à étudier le comportement externe des objets. Même le physicien, qui pourtant étudie le réel dans ses comportements les plus simples, doit bien à un certain moment considérer que son atome n'en est pas un et tenter de le voir dans sa complexité, c'est-à-dire étudier son organisation. Il construira alors une structure scientifique ou une théorie scientifique générale. Le point de vue qu'il adoptera sera macroscopique.

L'objet, vu dans sa complexité, devra être analysé en ses éléments constitutifs. Contrairement à la démarche précédente qui, ne prenant en compte qu'un seul type de comportement, considérait des objets tous similaires, l'approche macroscopique différenciera ces éléments en classes, chacune caractérisée par des comportements types. Ainsi le physicien atomiste distinguera-t-il des pro-

---

6. De telles tentatives ont pourtant été faites, notamment en économie, j'aurai l'occasion d'en parler plus loin. Des tentatives de théories restreintes intéressantes, mais au pouvoir explicatif désespérément bref sont celle de la bio-psychologie, qui tend à réduire tous les comportements du vivant à un principe de perpétuation génétique ou encore celle que je qualifierai de " socio-psychologie chimique " d'E. Schoffeniels, *L'anti-hasard*, Paris, 1973 (*cf.*, notamment, p. XII), qui tend à ramener tous les comportements humains, et jusqu'aux problèmes les plus fondamentaux des sociétés au jeu strictement biochimique des hormones et des phéromones. Cela peut bien sûr fournir une sorte d'explication au fait que Victor Hugo aimait les femmes et écrivait des poèmes, mais non au talent qu'il mettait à écrire. Et encore moins au mouvement romantique.

tons et des neutrons, des électrons et toute la gamme des particules qui interviennent dans le fonctionnement de l'atome, chaque catégorie étant dotée de
propriétés, de comportements caractéristiques. Une intuition produira, à partir
de cette analyse, une synthèse explicative et généralisante qui produira une
structure scientifique, soit un schéma capable de rendre compte de la complexité de l'objet. Tout autant que les principes qui fondent la théorie scientifique restreinte, la structure scientifique est abstraite ; elle est une matrice
logique dépourvue de coordination immédiate au réel[7].

FIGURE 2. LA STRUCTURE SCIENTIFIQUE.

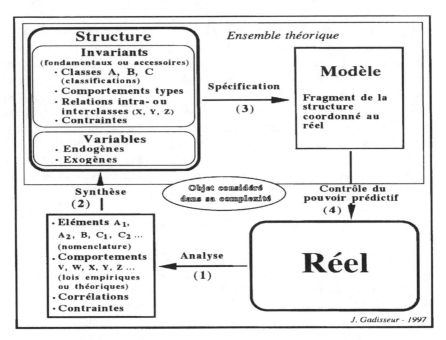

La structure scientifique, ou théorie scientifique au sens large, est une synthèse
généralisante et simplificatrice représentant la complexité interne de l'objet,
selon le point de vue macroscopique. Elle identifie les variables et les invariants
qui déterminent son fonctionnement, elle les répartit en classes, décrit leurs
comportements et leurs interactions. Elle est une matrice logique, abstraite, non
directement coordonnée au réel, dont cependant la spécification en modèles qui
en traduisent des fragments permet d'effectuer des prévisions sur le comportement du réel.

7. On comprend donc que la structure n'appartient pas à l'objet. La structure est un construit,
une représentation de l'objet dans sa complexité. Dire que le réel est structuré, c'est déjà le comprendre donc le représenter. Mais nous parlons toujours comme si les choses étaient les représentations que nous en formons. Il le faut bien, puisque ces représentations sont tout ce que nous en
connaissons.

La confrontation directe de la structure scientifique au réel est donc impossible. Il faudra en spécifier l'une ou l'autre partie et construire un modèle empirique dont on puisse déduire une prédiction sur le comportement du réel. La confrontation de la prédiction à l'observation permettra de délimiter l'étendue de validité de la structure théorique.

La structure comporte des variables qui en caractérisent l'état de fonctionnement — le terme étant pris ici dans un sens extrêmement général. Ces variables pourront être soit endogènes et résulter du fonctionnement de la structure, soit exogènes et consister alors en influences externes. De manière générale, ces variables comportent un domaine de variation admissible, en dehors duquel elles soumettent la structure à des tensions qui s'opposent à son fonctionnement normal et compromettent sa stabilité. La stabilité de la structure atomique est par exemple conditionnelle à la température ou à la vitesse. Ou bien, pour choisir un exemple à l'autre extrême, une structure sociale pourra être menacée par l'existence d'un chômage excessif, susceptible de la déstabiliser.

La structure est caractérisée par ses invariants. Ces invariants sont les classes d'éléments qui la composent, les relations intra- et interclasses qui s'établissent entre les éléments, et une série de contraintes ou de constantes qui pèsent sur les comportements des éléments et donc sur le fonctionnement de la structure. Qu'un seul de ces invariants se modifie, et toute la structure est qualitativement changée, c'est-à-dire que son comportement externe se trouve modifié. En outre, toute structure possède des invariants fondamentaux, soit un jeu d'invariants interdépendants tels que si l'un vient à changer, l'organisation qu'elle représente cesse d'exister. Elle se décompose — pour éventuellement se recomposer autrement. Il s'agira, à un bout de l'échelle, de fusion ou de fission nucléaire, à l'autre bout de révolution sociale, dans le sens d'une réorganisation complète des rapports sociaux, avec de nouvelles classes, de nouveaux comportements et jusqu'à de nouvelles mentalités[8].

La méthodologie qui préside à l'établissement des structures scientifiques est plus diverse et plus complexe que celle qui fonde la construction des théories restreintes. Encore une fois, le schéma par lequel je la représente n'est qu'une formalisation logico-pédagogique. Je veux dire que c'est ainsi qu'on peut expliquer les rapports qu'une théorie scientifique au sens large entretient avec le réel qu'elle tente de représenter. L'histoire de son invention est évidemment différente, faite d'approximations successives, de l'élaboration de modèles partiels qui, effectivement traduisent un fragment d'une structure, mais d'une structure dont justement on ne connaît encore que des fragments, et qu'il

---

8. Je suis bien conscient, passant ainsi de la structure du plus petit objet complexe que constitue l'atome à celle de l'ensemble le plus complexe qui soit — l'homme et ses sociétés —, d'imposer au lecteur un considérable et difficile effort de transition. Mais c'est que j'espère, faisant ainsi, montrer l'unité épistémologique des structures scientifiques. Nous n'avons pas un esprit si puissant que nous puissions varier beaucoup nos modes de représentation — seuls les contenus varient.

s'agira de recomposer, puis de compléter et de corriger au fur et à mesure que son progrès éclaire à la fois le réel et les lacunes qu'elle comporte.

Empruntons un exemple à la botanique. En physiologie végétale, on commence à bien connaître la structure de la plante et à élucider les uns après les autres les mécanismes de son fonctionnement. Mais il reste bien des mystères. Un de ceux-ci est le " facteur floral ". On sait que les plantes de régions tempérées et à floraison estivale, arrivées à un certain degré de maturité, réagissent à des variations de leur environnement — l'allongement de la durée du jour — et produisent des bourgeons floraux. Or, la relation qui s'établit entre la durée du jour — relation très déterminante, puisqu'une seule exposition à un jour long suffit — et la production de bourgeons floraux par le méristème reste inconnue. On suppose qu'il s'agit d'un processus vraisemblablement chimique ou hormonal, qu'on appelle le " facteur floral " en attendant d'en savoir plus. Mais on ne sait pas quelle partie de la plante le produit ni comment ; on ne sait pas non plus comment il exerce une influence sur le fonctionnement d'une autre partie de la plante. C'est-à-dire qu'il reste des éléments de la structure à distinguer et qu'il reste à en préciser les relations avec les cellules qui donneront les bourgeons floraux. Pour dire les choses en termes épistémologiques, on a dans ce cas établi la loi microscopique de la floraison : la plante fleurit après avoir été exposée à une longue période de lumière, ce qui décrit bien son comportement externe, mais on ne peut pas encore expliquer ce comportement particulier par la structure scientifique.

C'est dire que la méthode expérimentale joue un très large rôle dans la construction des structures scientifiques. Mais il s'agit d'une méthode expérimentale d'une autre ampleur que celle envisagée par Popper[9]. C'est-à-dire qu'il s'agit d'une méthode fondée sur l'observation des expériences afin de produire des hypothèses acceptables, plutôt qu'afin de réfuter des hypothèses préexistantes : les tables de Francis Bacon, car il s'agit d'identifier des corrélations.

___

9. Je pense qu'on a dit à propos de vérification ou de falsification beaucoup de sottises. On s'est surtout trompé de sens : ce ne sont pas les hypothèses ou les propositions scientifiques que l'on vérifie par l'expérience, mais les observations, et la méthode expérimentale est essentiellement une méthode de critique des observations, afin d'en assurer la signification. Pour le reste, il ne s'agit que de préciser les limites de validité des constructions théoriques, ou alors d'opérations plutôt politico-médiatiques, comme dans le cas du pendule de Foucault. D'ailleurs, les choses sont loin d'être aussi simples que le croyait Popper. Reprenons l'exemple du facteur floral. L'hypothèse pourrait être : " Il s'agit d'une hormone ". On va donc cultiver des plantes, provoquer leur floraison puis opérer des prélèvements et effectuer des analyses sur des groupes de plantes à différents stades du processus de formation des bourgeons floraux. Supposons qu'on n'identifie aucune hormone. Cela signifie-t-il qu'il convient de rejeter l'hypothèse ? Évidemment non. On ne pourra la rejeter que si l'on identifie positivement le facteur floral et que l'on constate qu'il ne s'agit pas d'une hormone. Mais alors, on aura fait bien plus que rejeter l'hypothèse hormonale : on en aura accepté une autre à laquelle on n'avait peut-être pas d'abord songé. En fait, la méthode expérimentale sert à produire les propositions scientifiques, non à les falsifier. La science n'est pas un tribunal. Si c'était le cas, on n'aurait jamais rien appris. J'ajouterai encore ceci : non seulement toute science comporte des propositions infalsifiables — précisément les axiomes et les définitions qui en sont l'indispensable fondement —, mais encore la philosophie et la mathématique ne comportent que des propositions infalsifiables. Elles sont pourtant bien des sciences.

L'EXPLICATION SCIENTIFIQUE

Ce qui vient d'être dit montre que l'explication scientifique a clairement une double nature. D'une part, dans le cas des théories restreintes, l'explication des lois réside dans les principes dont elles sont déduites. Le principe de la gravitation universelle et celui de l'inertie expliquent la constante géodésique. Mais on comprend que l'explication soit ici purement abstraite et donc limitée : on explique une proposition particulière par une proposition plus générale dont elle dérive. La cause évoquée est donc une cause logique, soit un changement de niveau dans la hiérarchie des propositions. Je puis dire que les corps tombent à peu près vers le centre de la Terre en raison de la gravitation universelle. Mais je ne puis pas dire comment il se fait qu'il y ait une gravitation universelle[10].

L'explication scientifique proprement dite, celle du " comment " n'est donnée que par la structure scientifique, qui rend compte de la complexité du réel. Dire : " La plante fleurit lorsque les jours sont longs ", c'est constater une régularité de comportement, et donc une corrélation. C'est déjà beaucoup, puisque la prévision et l'action sont rendues possibles par cette connaissance toute simple. Mais il n'est guère utile ni intéressant " d'expliquer " — d'ailleurs très logiquement — ce comportement par le principe général que la plante fleurit au moment où elle va disposer d'un maximum d'énergie lumineuse pour opérer la photosynthèse. L'explication véritable, c'est-à-dire la description du fonctionnement de la plante qui conduit à la floraison, réside dans la structure scientifique que l'on construit à propos d'elle. Seule cette description permet en effet un contrôle effectif de la floraison. L'homme est *praxis*, et la science qu'il forme doit posséder une effectivité, sinon elle n'est que forme dépourvue de substance.

Une autre façon de dire, c'est que l'explication scientifique doit mettre en lumière des causes réelles. Alors seulement, la connaissance est véritable connaissance, puisque la causalité réelle implique une possibilité d'action.

---

10. Pour fournir une véritable explication, il faudrait que l'on puisse construire une structure scientifique capable de représenter le fonctionnement intime de la matière et de montrer comment, par ce fonctionnement, la matière produit un comportement externe qu'on appelle la gravitation. Je sais que tout physicien rêve d'une théorie unifiée des champs, même si seulement quelques-uns le confessent. Je pense aussi que beaucoup considèrent comme évident qu'une telle théorie devrait être une théorie scientifique restreinte. Peut-être se trompent-ils et peut-être l'explication des différentes interactions, réside-t-elle justement dans une structure scientifique. Quitte à faire sourire, allons jusqu'au bout — puisque je ne suis pas physicien, il y a des audaces que je puis me permettre : construire une structure qui rende compte du comportement des particules élémentaires supposerait qu'on arrive à les décomposer en éléments plus petits. La chose n'est peut-être pas impossible, si l'on accepte l'idée qu'à ce niveau l'élément constitutif peut être d'une nature différente et que le concept même de dimension perd sa pertinence.

Les relations entre théorie scientifique restreinte et structure scientifique

Je l'ai déjà dit, les approches microscopique et macroscopique sont indissociables et alternatives. Les lois, ou régularités de comportement mises en lumière par l'approche microscopique, d'une part sont expliquées par l'approche macroscopique de l'objet, d'autre part sont mises à profit par la structure scientifique décrivant l'ensemble complexe dont l'objet est un élément.

Il faut remarquer cependant que, dans la relation théorie restreinte-structure scientifique, la réduction axiomatique des lois ne joue de rôle que dans la mesure où elle a permis de compléter et de préciser la description du fonctionnement externe des objets. C'est-à-dire que la structure scientifique n'utilise jamais que la simple description des comportements, indépendamment de l'explication logique que la théorie restreinte leur apporte. Une structure géologique tiendra bien sûr compte de l'attraction terrestre ou de la loi des marées, mais ne se préoccupera pas de justifier ces lois en termes de gravitation universelle ou de théorie de la relativité — seuls les comportements effectifs comptent. Qu'ils aient été simplement observés — lois empiriques — ou déduits des axiomes — lois théoriques — n'a guère d'importance du point de vue macroscopique.

Cependant, si la structure scientifique peut fort bien se dispenser de l'apport des théories scientifiques restreintes, l'inverse n'est pas vrai. Dès que la théorie scientifique restreinte est confrontée au réel, elle doit impérativement se couler en forme de structure, ou si l'on préfère en un modèle qui reflète une structure. La remarque est d'une extrême importance.

La chose, si on prend la peine d'y réfléchir un peu, va absolument de soi : ne serait-ce que pour observer, il faut au moins qu'il y ait deux choses. Même les comportements les plus simples de la matière la plus élémentaire ne peuvent s'exprimer que s'il existe un environnement, qui évidemment comporte au moins une autre chose. Puisqu'il y a une pluralité de choses, il y a un degré de complexité dont seule une structure est capable de rendre compte. C'est ainsi que la théorie de la gravitation universelle suppose qu'il existe au moins deux corps, que la thermodynamique suppose l'existence d'une source et d'un flux, donc d'un degré de complexité du réel. C'est-à-dire que la théorie scientifique restreinte, pour s'exprimer, a besoin d'une forme structurelle.

Maintenant, cette forme structurelle ne correspond évidemment pas à la complexité d'un ensemble réel, sauf s'il faut prévoir le résultat d'une expérience concrète — il faudra alors préciser le mieux possible la " structure " de l'expérience ; c'est-à-dire, suivant la terminologie que j'ai adoptée ici. spécifier un modèle empirique, reflet d'un fragment d'une structure théorique. Ainsi, vérifier la loi de la chute des corps suppose l'organisation d'une expérience qui tienne compte du frottement de l'air, suivant sa pression et sa température, de la latitude, de l'éventuelle présence de mascons, etc.

LA STRUCTURE-ZÉRO

Indépendamment du modèle empirique, l'expression des conséquences d'une théorie scientifique restreinte nécessite toujours la construction d'une structure imaginaire. Par exemple un système mécanique à deux, à trois ou à n corps. Il ne sera pas nécessaire, puisqu'il ne s'agit pas d'observer ni d'interpréter les résultats d'une expérience, de spécifier ni de paramétriser cette structure. D'autre part, on la choisira évidemment la plus simple possible, et donc dépourvue d'organisation. Pour cette raison, j'appelle cette structure une *structure-zéro*.

Puisqu'il s'agit de théorie scientifique restreinte, et donc d'un seul type de comportement appartenant à une seule catégorie d'objets, il ne peut évidemment y avoir d'organisation dans la structure imaginaire — structure théorique, dans le sens d'hypothétique. Supposer une organisation, c'est évidemment supposer une variété d'éléments et de comportements, c'est supposer entre les éléments des relations différenciées, donc perturbatrices par rapport au comportement unique qu'on se propose d'étudier.

La structure-zéro, dépourvue de toute organisation et qui n'a pour complexité que le seul nombre des éléments qu'elle comporte, est le lieu de transition entre la structure et la théorie scientifique restreinte. Elle est une anamorphose de la structure en théorie restreinte.

Les physiciens sont familiarisés avec de telles structures. Tous les systèmes gravitationnels relèvent de cette catégorie. Le souci d'assurer l'indifférenciation des relations entre les éléments est poussé si loin que l'on suppose parfois avoir affaire à des GSP — Grands Systèmes de Poincaré —, soit à des systèmes gravitationnels tellement vastes que les phénomènes de bordure deviennent négligeables. En effet, les corps situés en périphérie se trouvent avec les autres dans une relation particulière, les attractions auxquelles ils sont soumis étant dissymétriques, de sorte que la simple limitation du système lui confère un degré d'organisation, sans doute faible, mais fort gênant du point de vue de la déduction théorique.

Les structures-zéro — qu'on appelle souvent systèmes ou modèles théoriques — sont nombreuses. Les gaz parfaits sont des structures-zéro au sens que je viens de définir, comme d'ailleurs en général les états amorphes de la matière. La masse critique des physiciens nucléaristes est une structure-zéro, le nuage de Jeans en est une autre comme certains modèles en thermodynamique ou les mélanges homogènes des chimistes.

On comprend bien l'utilité de pareilles structures. Elle est d'abord de tester la puissance prédictive des instruments théoriques, comme dans le cas des modèles gravitationnels à trois corps et plus, pour lesquels il apparaît que, les équations n'étant pas intégrables, la prévision exacte des mouvements relatifs est impossible, à moins qu'en terme de probabilités. Elle est ensuite d'explorer logiquement les conséquences d'interactions simples entre un grand nombre d'éléments, et notamment de vérifier, toujours au niveau de la prédiction logique, la stabilité de la structure-zéro. Le modèle de Jeans par exemple, montre qu'un nuage de matière uniformément distribué dans l'espace génère des effondrements gravitationnels susceptibles de donner naissance aux galaxies et aux astres qui les composent, et que donc la structure se différencie et gagne en organisation. Ou bien, la masse de matière fissile des physiciens nucléaires montre qu'il existe une frontière de stabilité, au-delà de laquelle apparaît un comportement qui provoque la destruction de la structure-zéro. Indéniablement, la structure-zéro est un instrument méthodologique à la fois essentiel et puissant chaque fois qu'il s'agit d'étudier les conséquences d'un comportement simple.

La structure-zéro ne permet évidemment pas de rendre compte de la complexité du réel. Néanmoins, on trouve assez curieusement des structures-zéro censées représenter l'objet le plus complexe qui soit : la société humaine.

LE MODÈLE ÉCONOMIQUE NÉO-CLASSIQUE : UNE STRUCTURE-ZÉRO

Il me faut revenir au début de mon propos. J'ai d'abord énoncé que la focalisation du monde scientifique sur les constructions de la physique, et précisément sur les théories restreintes qu'elle seule peut élaborer, parce que son objet est le plus simple qui soit — je puis ajouter maintenant : et sur les structures-zéro qui en sont le prolongement —, a gravement entravé le développement des sciences humaines.

Depuis maintenant bien plus d'un siècle, la science économique s'accroche obstinément à une représentation de la société des hommes qui n'est rien d'autre qu'une structure-zéro construite à partir d'une théorie restreinte censée expliquer et décrire le comportement de l'*homo economicus*. Il s'agit du modèle néo-classique de l'économie de marché, ou modèle walrassien. Le contenu de la théorie scientifique restreinte qu'est la théorie économique néo-classique est le suivant, y compris les hypothèses accompagnatrices :

1°   L'*homo economicus* a des besoins globalement illimités, c'est-à-dire que les biens sont toujours relativement rares, ce qui est la condition pour qu'ils aient un prix non nul, en somme pour qu'ils soient des biens économiques.

2°   L'*homo economicus* est rationnel et hédoniste, c'est-à-dire qu'il cherche à maximiser sa satisfaction. Néanmoins, la satisfaction qu'il éprouve à posséder[11] un bien tend à diminuer à mesure qu'il en possède davantage.

3°   L'*homo economicus* a des préférences et il peut les ordonner (sinon il ne pourrait exercer sa rationalité).

4°   Il est propriétaire de quantités déterminées d'un certain nombre de biens (ce sont ses dotations initiales).

5°   Les règles de l'honnêteté sont respectées, tant vis-à-vis de la propriété que de la sincérité ou du respect des contrats.

6°   L'*homo economicus* est parfaitement au courant des intentions de tous les autres (c'est la transparence du marché).

7°   Les conditions sont telles qu'aucun intervenant n'a intérêt à modifier son offre ou sa demande d'un bien quelconque afin d'en faire changer le prix (c'est la condition d'atomisticité du marché).

8°   Les prix de marché, soit les rapports de quantité entre les biens échangés, sont uniques et règlent tous les contrats.

9°   Tous les marchés fonctionnent simultanément et instantanément (il n'y a donc pas de processus d'ajustement).

10° Certains biens — appelés " facteurs " — peuvent être transformés en d'autres biens au travers de fonctions de production (des " firmes " dépourvues d'existence réelle, de matérialité : ce sont des fictions opérationnelles) qui ont pour caractéristique de ne pas comporter d'économie d'échelle (le coût de la transformation augmente proportionnellement ou plus que proportionnellement aux quantités produites).

Moyennant ces conditions, qui décrivent une structure-zéro composée d'un certain nombre d'hommes uniquement préoccupés d'opérer entre eux des échanges afin de maximiser leur satisfaction, et de " firmes " — objectivement serviables, quoique animée par l'intention de maximiser leur profit, mais qui n'en réalisent pas le moindre — qui opèrent les transformations les plus adéquates, on détermine un équilibre, caractérisé pour chaque bien par un doublet " quantité totale échangée-prix de l'échange ". Cet équilibre est la solution d'un système d'équations différentiels[12] et simultanées qui intègrent les offres

11. La consommation proprement dite — soit la destruction de la valeur des biens — n'est pas prise en compte par le modèle néo-classique. Il s'agit d'une décision ultérieure. Le bien acquis par l'agent sera consommé ou conservé, cela n'importe pas pour le fonctionnement du marché. Sinon, la principe de la conservation de la valeur ne serait pas respecté.

12. Qu'il s'agisse bien d'une structure-zéro est encore attesté par le fait que le modèle économique néo-classique se heurte à la même difficulté méthodologique que celle rencontrée dans les systèmes gravitationnels à trois corps et plus. En effet, lorsqu'il y a trois biens ou davantage, les équations cessent d'être intégrables en général. Il y a déjà longtemps que les économistes ont résolu ce problème théorique en faisant intervenir les probabilités — l'incertitude. I. Prigogine, *La fin des certitudes*, a récemment préconisé le recours à un procédé analogue pour résoudre le problème de Poincaré.

et les demandes de chacun des intervenants pour chacun des biens et pour tous les prix, compte tenu des fonctions de production qui décrivent les possibilités de transformation des biens en d'autres biens.

Cet équilibre comporte des caractéristiques remarquables[13] :

1°   Les quantités échangées sont maximum, de même que les quantités produites (transformées), compte tenu bien sûr des conditions de départ qui sont les dotations initiales, les préférences et les fonctions de production.

2°   En conséquence, on est au plein emploi des facteurs de production (ce sont les biens qui peuvent être transformés en d'autres biens). Ce qui signifie que tous ceux qui disposaient de facteurs de production et ne les ont pas vendus les ont conservés parce que le prix d'équilibre du marché leur paraissait insuffisant. Cela signifie notamment que s'il existe un chômage, il est volontaire.

3°   La satisfaction est maximum, compte tenu des conditions. En d'autres termes, chacun obtient une satisfaction qui est la plus grande possible étant donné les dotations dont il disposait, ses propres préférences et l'état du marché, c'est-à-dire les prix. D'où on déduit que la répartition des biens est optimale.

4°   Donc, à l'équilibre, plus aucun échange n'est souhaité par personne. On dit que les marchés sont vidés.

Le modèle néo-classique de l'économie de marché ressemble beaucoup à un modèle thermodynamique auquel on aurait apporté quelques sophistications : un nombre indéterminé de biens, des fonctions de transformation non linéaires, l'absence de transfert sans exacte contrepartie, des taux de conversion (des prix d'échange) variables et une simplification : l'instantanéité des processus et l'abolition de la contrainte d'espace. Il est remarquable en outre que ce modèle comporte, comme en thermodynamique, un équilibre stable, c'est-à-dire un état dans lequel les échanges sont devenus impossibles. La différence est qu'en thermodynamique, cet état n'est atteint qu'asymptotiquement, tandis qu'en économie, il l'est instantanément. Il est remarquable aussi que, de la même manière que l'entropie est maximum à l'équilibre thermodynamique, la satisfaction — ou l'utilité subjectivement appréciée — le soit à l'équilibre économique. Enfin, comme en physique où il n'y a ni création ni destruction globale d'énergie, le modèle néo-classique n'envisage ni la création, ni la destruction de la valeur. L'échange s'opère évidemment à valeur égale, de même que la transformation des biens en d'autres biens : les intrants de la firme ont exactement la même valeur que les extrants. La loi de la conservation de l'énergie a donc pour pendant implicite celle de la conservation de la valeur.

13. Conclusions scientifiquement triviales cependant, n'en déplaise à mes confrères économistes, mais idéologiquement fort pondéreuses.

Ce modèle néo-classique est très évidemment une structure-zéro, expression élémentaire, quoique mathématiquement complexe, d'une théorie scientifique restreinte. Il me paraît clair qu'on s'est trompé de théorie, et qu'on a voulu représenter l'organisation d'un objet extrêmement complexe — la société économique — par une construction qui ne peut rendre compte que des comportements simples d'objets indifférenciés.

Si la structure-zéro de l'économie néo-classique pouvait avoir un intérêt autre que formel, rhétorique ou idéologique, ce serait de démontrer par exemple que cette structure-zéro, représentant une économie parfaitement démocratique, sinon égalitaire, peut être stable, susceptible de conserver les caractéristiques que lui confèrent les hypothèses sur lesquelles on la fonde, ou au contraire que son fonctionnement aboutit à les transgresser. Curieusement, les économistes, qui pourtant se sont attachés à faire fonctionner leur modèle sous des conditions qui ne respectent pas parfaitement l'une ou l'autre des hypothèses, et qui sont arrivés à montrer qu'alors un équilibre se réalise néanmoins, qui certes n'est pas le meilleur, mais est le *second best*, compte tenu des conditions, ne se sont jamais attachés à simuler une série de fonctionnements successifs en répercutant de l'un à l'autre les modifications qui en résultent pour les dotations initiales. L'auraient-ils fait qu'ils auraient inévitablement constaté que, tout comme il se produit des effondrements gravitationnels dans l'espace de Jeans, il doit se produire des accumulations chrématistiques sur le marché de Walras : des montagnes de richesse séparées par des gouffres de pauvreté — et en conséquence une re-partition qualitative de la société entre propriétaires et prolétaires[14].

Il est bien sûr arrivé aux économistes néo-classiques de prendre en compte des fonctionnements successifs du modèle, afin notamment d'expliquer la croissance économique par l'accumulation du capital. Bizarrement, ils ont cependant toujours fait semblant que la croissance du capital était indifférente relativement à la répartition des biens. Exactement comme si un esprit égalitaire avait, entre chaque partie soigneusement battu et redistribué les cartes, alors même que leur nombre et leur valeur augmentaient. En somme, on peut dire que les économistes se sont tellement attachés à la représentation réductionniste qu'ils formaient de leur objet, qu'ils ont radicalement refusé même d'admettre que la structure-zéro qu'ils avaient construite fût instable, et capable de se transformer, par son seul fonctionnement, en une structure vraie, c'est-à-dire comportant des différenciations qualitatives, génératrices de véritable organisation.

14. Ou même aller plus loin dans l'analogie avec les constructions de la thermodynamique, en considérant que les flux économiques passent par des structures dissipatives capables de générer des " néguentropies ". Soit un fonctionnement au terme duquel la répartition de la valeur serait plus inégale relativement aux préférences. Mais ce serait bien sûr encore insuffisant pour rendre compte du fonctionnement réel de la société économique dans toute sa complexité.

Dans les perspectives que je viens d'ébaucher, il apparaît bien sûr que le modèle économique néo-classique est inadéquat, et donc à peine scientifique[15]. La société économique est hautement organisée, les hommes qui la composent sont nettement différenciés quant à leurs comportements, elle fonctionne dans un temps historique et les attitudes économiques ne sont nullement séparables des autres aspects de l'activité humaine : le politique, le démographique, le religieux, le gnoséologique… Il convient dès lors évidemment de construire à propos de la société économique, à supposer même que quelque chose de tel soit analysable distinctement, une structure scientifique proprement dite.

A condition d'adopter les formes de représentation qui conviennent et les modes d'explication adéquats, tout le réel, y compris l'homme, ses sociétés et leur histoire est scientifiquement intelligible, et suivant les mêmes schèmes épistémologiques que la biologie, la géologie, la chimie organique ou la physique atomique ; donc, il n'y a pas de domaine qui soit plus ou moins scientifique qu'un autre, ni même de science qui soit par la nature de son objet plus ou moins exacte qu'une autre.

Il importe seulement de se convaincre que la science véritable comporte bien plus et bien mieux que la seule réduction axiomatique des lois qui décrivent le comportement externe des choses.

---

15. Malgré la révision keynésienne ou le schisme schumpétérien. On ne pourrait même pas — et on ne penserait certainement pas à le faire — appliquer le modèle néo-classique à une fourmilière, trop évidemment organisée. On prétend pourtant rendre compte grâce à lui de cet objet encore bien plus complexe et plus hautement organisé qu'est la fourmilière humaine.

POLARITIES WITHIN MATHEMATICS[1]

Roman DUDA

Mathematics is a particular field of human activity. Since long it has attracted attention of not only mathematicians but also of many philosophers, including the greatest ones. One of them has described its role in our culture as follows : " I will not go so far as to say that to construct a history of thought without profound study of the mathematical ideas of successive epochs is like omitting Hamlet from the play which is named after him. That would be claiming too much. But it is certainly analogous to cutting out the part of Ophelia. This simile is singularly exact. For Ophelia is quite essential to the play, she is very charming — and a little mad "[2].

In its several thousand years long history, mathematics has served several lords and pursued different goals, some of which are openly opposite. Nevertheless, most of them are still valid and remain attractive. This phenomenon allows us to look upon mathematics as that discipline which is spread out between several polarities within which each pole attracts it in a certain way. One can say that each pole determines attitudes and aims of some mathematicians, but never all of them, and even one and the same mathematician can assume various attitudes towards them during his/her life time. Submerged in that field of many tensions, mathematics maintains a creative balance.

The polarities which I have in mind are the following :

Platonism - empiricism

the finite - the infinite

the discrete - the continuous

the approximate - the exact

certitude - probability

---

1. A preliminary version of this article appeared as : R. Duda, " Mathematics : Essential Tensions ", *Foundations of Science*, 2 (1997), 11-19.

2. A.N. Whitehead, *Science and the Modern World*, 1960, 26.

simplicity - complexity

load of the past - shortcuts to research frontiers

ugliness - beauty

the pure - the applied

unity - multiplicity

The aim of this paper is to describe each polarity briefly. I will also allow myself some comments and questions.

REALISM - IDEALISM

According to Platonists, mathematical objects enjoy real existence, independent of human thought. On the other hand, opponents to Platonism claim that mathematical objects are human artefacts and come mostly by abstraction from real objects in the world around us. Although some opponents allow also independent inventions (but still denying their real existence), let me call them summarily empiricists. Both attitudes, Platonism and empiricism, are well known since antiquity. The father of Platonists is Plato himself. Empiricism, in the sense described above, is as old and the first man who has consciously took that attitude was Aristotle[3]. Both attitudes have some variants and each claim many supporters[4].

Leaving aside subtle philosophical differentiations, let us proceed to the questions which are asked by many and which reveal, at least partly, the essence of the considered polarity.

What does mathematics depict : the actual physical world, which surrounds us (assuming that it exists), or the world of ideal entities existing somewhere beyond it, or something different again ?

If mathematical objects are ideal entities, where does this world of the invisible exist ? And how does it interact with the physical world ? In particular, why is then mathematics so effective in describing the physical reality[5] ? How could we attain knowledge of mathematical reality which transcends space and time ?

But if mathematical objects are abstract forms of matter, what is the relation of such a form to the matter itself ?

The problem hidden behind the polarity Platonism - empiricism can also be put in the form of the question : is mathematics a discovery or a creation ? If

---

3. *Cf.* Aristotle, *Physics*, Book II, 193b ; *Idem*, *Metaphysics*, Book K, 1061, and Book M, 1077b. See also B. Russell, *History of Western Philosophy*, London, 1957.

4. *Cf.* P. Kitcher, *The Nature of Mathematical Knowledge*, New York, 1983.

5. E. Wigner, " The Unreasonable Effectiveness of Mathematics in the Natural Sciences ", *Communications in Pure and Applied Mathematics*, 13 (1960), 1-14.

discovery, what are we discovering ? And if creation, is there any " logic " of these acts of creation ?

Let us mention a spectacular example the following theorem of Euler linking together basic constants of mathematics : $e^{\pi i} + 1 = 0$.

Was Euler a discoverer of an already existing fact or an inventor of a new construction ?

What are then the objects of our study ? What is number, point, sphere, Riemann zeta function, Lebesgue integral, or Haar measure ? Is there any similarity between a mathematician and a geographer or a biologist ?

Responding to those and similar questions, mathematicians are strongly divided. The popular anecdote says that the typical working mathematician is a Platonist on weekdays but opposes Platonism while on Sunday leisure[6].

THE FINITE - THE INFINITE

The concept of infinity appeared among the early Greeks, and at once, in the aporias of Zeno of Elea, revealed its paradoxical character : if infinity is not admitted, then Achilles is not able to catch up with the tortoise, and an arrow in flight is at rest.

Aristotle drew a distinction between the potential infinite and the actual infinite[7]. The potential infinite is that whose infinitude exists over time, in any process that can never end, e.g. in counting. For Aristotle it is a fundamental feature of reality. The actual infinity, on the other hand, is that whose infinitude is given at some point in time, all at once. Aristotle insisted that all objections to the infinite concern the actual infinity. The Greeks followed Aristotle, thus allowing the potential but not the actual infinity. For instance, Euclid's theorem that the sequence of prime numbers is infinite runs as follows : prime numbers are more than any assigned multitude of prime numbers[8].

B. Bolzano had written a book on the topic[9] and Cantor's set theory, dealing with the actual infinity and presenting its coherent, rigorous, and systematic treatment, was met both with a warm welcome (Hilbert) and with much resistance (L. Kronecker, H. Poincaré)[10]. Can we actually tame the truly infinite ?[11].

6. *Cf.* Ph.J. Davis, R. Hersh, *The Mathematical Experience*, Boston, 1980, 321-322.

7. Aristotle, *Metaphysics*, 204a, 206a ; *Idem*, *Physics*, 994b.

8. Euclid, *Elements*, Book IX, Proposition 20. Quoted after Th. Heath, *The Thirteen Books of Euclid's Elements*, Dover, 1956 (3 vols).

9. B. Bolzano, *Paradoxes of Infinity*, Translated by D. Steel, New Haven, 1950.

10. H. Poincaré, *Science and Method,* Dover, New York, 1952. See also G.H. Moore, *Zermelo's Axiom of Choice, Its Origins, Development and Influence*, Springer, 1982.

11. A. Moore argues that we can't, see his *The Infinite*, London, 1990.

Mathematicians are divided. Some accept the actual infinity unreservedly, while other seem to concur in an opinion that since the physical world is basically finite (although of unimaginable dimensions), the infinite can be admitted only as a tool. And although nobody is expelling us from the paradise of the infinite (Hilbert), it seems that mathematics itself has been changing the mood by advancing nowadays its finite branches more vividly than the infinite ones.

THE DISCRETE - THE CONTINUOUS

There is a correlation between this polarity and the former one. However, they are not identical, e.g. discrete structures can be either finite or infinite.

Polarity the discrete — the continuous has been first observed by Pythagoreans, who distinguished between a discrete part of mathematics (consisting of arithmetic studying absolute discrete quantities and of music dealing with relative discrete quantities) and a continuous part of mathematics (consisting of astronomy studying variable continuous quantities and of geometry dealing with constant continuous quantities)[12]. And from that time on geometry and astronomy were dominant in Greece[13], whereas in the modern times it is analysis that took the lead.

Basic concepts of analysis are number, function, continuity. Number seems close to the discrete pole of the polarity but function and continuity do not. Nevertheless, the process of rigorisation of analysis in 19th century[14] has pushed it in the direction of the discrete (*cf.* concepts of Cauchy and of Weierstrass). And nowadays it is the discrete part of mathematics that has been rapidly developed, e.g. the theory of graphs, computer science. One may risk an opinion that the centre of gravity of modern mathematics is shifting towards the discrete. Is it just a result of fruitful applications or an impact of modern technology (PC's, digital recording, etc.) or even, possibly, of some deeper reasons ? But even admitting that shift we feel justified to say that the continuous still remains very strong.

For Greeks, geometric figures having a measure (length, area, volume) could not be composed of non-measurable objects such as points, and therefore the differentiation between the discrete and the continuous was much sharper at that time. To what extent did the concept of function as a fundamental math-

12. *Cf.* H.W. Turnbull, *The Great Mathematicians*, Reprinted in the book : *The World of Mathematics, A Small Library of the Literature of Mathematics form A'h-mose the Scribe to Albert Einstein, Presented with Commentaries and Notes by James R. Newman*, vol. I, London, 1961, 73-168 (4 vols).

13. The influence of Euclid's *Elements* and of Ptolemaeus' *Almagest* was enormous and lasted until modern times.

14. *Cf.* M. Kline, *Mathematical Thought from Ancient to Modern Times*, New York, 1972, chapter 40.

ematical object reduce that polarisation ? How did the 19th century concept of a geometrical figure as a set of points contribute to that ?

THE APPROXIMATE - THE EXACT

The polarity can be illustrated by the search for solutions of equations. On the one hand, we can seek to show that they exist, while on the other hand, we can look for their numerical value, frequently approximate only.

This leads to a distinction between dialectic and algorithmic mathematics, which Henrici characterized as follows : " Dialectic mathematics is a rigorously logical science, where statements are either true or false, and where objects with specified properties either do or do not exist. Algorithmic mathematics is a tool for solving problems. Here we are concerned not only with the existence of a mathematical object, but also with the credentials of its existence. Dialectic mathematics is an intellectual game played according to rules about which there is a high degree of consensus. The rules of the game of algorithmic mathematics may vary according to the urgency of the problem on hand. We never could have a man on the moon if we had insisted that the trajectories should be computed with dialectic rigor. The rules may also vary according to the computing equipment available "[15].

Are we practitioners of algorithmic or dialectic mathematics ?

CERTITUDE - PROBABILITY

For a long time mathematics supported the deterministic understanding of the physical world by offering methodology which leads to results which are certain. Even if they were approximate, like the decimal representation of an irrational number, we knew the order of approximation and could improve it.

With the introduction of the concept of probability (at the beginning it was an innocent interest in card games) the picture was dramatically changed. Through mathematics the probability entered physics and other sciences. And original determinism had given way to a more probabilistic understanding.

Can probability be treated as a consequence of our ignorance[16] ? Or, perhaps, the probabilistic character of the physical world is intrinsic, and deterministic theories can only be regarded as a kind of approximation of our knowledge about it ?

Nowadays probability in mathematics is most often treated as a " probabilistic " measure on a specific field and thus probability theory has become a part of " good " mathematics. It seems, however, that the split remains. In other

15. Quoted after Ph.J. Davis, R. Hersh, *The Mathematical Experience, op. cit.*, 183.
16. Such was an understanding of Laplace, see his *Essai sur la probabilité.*

words, the choice of the way in the mathematical description of an object — deterministic or indeterministic — depends on the complexity of the model, on our knowledge, and on the practical possibilities.

SIMPLICITY - COMPLEXITY

Mathematics exists in human minds. To persist, it must be thought over and over again, it must be understood and memorised, it must be transferred in an ordered manner to other people. Inherent in that process is the tendency towards simplification (psychologists could perhaps say a word about it).

Consequently, mathematics changes continually. Axioms and concepts, theorems and proofs, mathematical procedures and theories are subjected to permanent reflection, resulting in their maximal purification of insignificant layers and in clarity and simplicity.

Good example of such a process is the concept of group which appeared in the mid-18th century as a set of some transformations of coefficients of a polynomial equation, turned out to be effective and interesting, and in the evolution nearly two centuries long has been simplified up to the present understanding of a set with some rules. Examples of that kind are many.

On the other hand, however, mathematical objects are becoming more and more sophisticated. Out of the already clearly transparent concept of algebraic group and out of the similarly simple concept of topological space there arose the concept of topological group, which has become a subject of still engaging intensive research. Differential manifold is perhaps even a more spectacular example. It comprises a geometrical structure (an atlas), an analytical structure (smooth charts), and algebraic structures (external forms, de Rham cohomologies, etc.).

Without simplification we would be drowned in the growing volume of hardly understandable stuff, and without complexification we would stop at the point of banality. Thus the two tendencies, apparently opposite, are in fact rather complementary.

LOAD OF THE PAST - SHORTCUTS TO RESEARCH FRONTIERS

It is sometimes argued that everything in mathematics remains true for ever. In fact, no mathematical position, if only properly formulated and/or demonstrated, can be falsified. Each year thousands of new theorems are added. In that way the volume of mathematics would grow very fast, soon becoming an impenetrable labyrinth of ideas, definitions, theorems, etc., with a danger of loosing its entity and usefulness.

So far, however, mathematics has retained its identity and the volume of mathematics, although enormous, still has human scale. This astonishing fact is due to two processes : simplification, which was discussed above, and selection, which consists in casting out unprofitable stuff. Casting out brings about a risk, *cf.* Ulam's dilemma[17], but it is unavoidable. Both processes lead together to a new organisation of mathematics, as witnessed, e.g., by subsequent editions of mathematical encyclopaedias.

The load of the past, however, remains enormous and thus creates the problem, whether progress in mathematics is possible. To make progress, mathematics needs researchers, but to do efficient research, one does not need to possess full knowledge of the past. In fact, such a knowledge would rather be an obstacle, constituting an unwanted burden for a researcher and restricting his/her inventiveness[18].

This is a real problem, and in the present days one may observe a growing gap between mathematicians oriented more to the burden of the past and those more to the exploration of new frontiers. The former remind us of custodians, the latter are rather like explorers. Although the share of custodians will probably grow, both mentalities are necessarily complementary and neither can disappear.

UGLINESS - BEAUTY

Each mathematician is aware of the important role of beauty as a guidance in research and as a stimulation in efforts to achieve simplicity. G.H. Hardy remarked once that : " Mathematical patterns like those of the painters or the poets must be beautiful. The ideas, like the colors or the words must fit together in a harmonious way. Beauty is the first test. There is no permanent place for ugly mathematics "[19].

According to S. Weinberg, the beauty of the perfect structure is economical, classic, resembling the order of the Greek tragedies[20]. Sense of beauty appears to be a useful strategy also in applications of mathematics to physics[21].

To offer one example, recall the Riemannian geometry, which was developed as a pure theory and became a divine gift for Einstein's theory of relativity.

THE PURE - THE APPLIED

Is mathematics a " pure " science, *i.e.* the pure work of mind, almost without any interference to the physical world ? Or should mathematics seek a jus-

17. *Cf.* Ph.J. Davis, R. Hersh, *The Mathematical Experience, op. cit.*, 20 ff.

18. To deal with that phenomenon, G. Bachelard has come to the concept of an epistemological obstacle, see his *La formation de l'esprit scientifique*, Paris, 1989.

19. G.H. Hardy, *A Mathematician's Apology*. Quoted after S. Weinberg, *Dreams of a Final Theory*, New York, 1992.

20. In the book S. Weinberg, *Dreams of a Final Theory, op. cit.*, the author formulates three hypotheses explaining the mysterious efficiency of a beauty criterion in mathematics.

21. F. Le Lionnais, " La beauté en mathématiques ", in the book : F. Le Lionnais (ed.), *Les Grands courants de la pensée mathématique*, Paris, 1948, 437-465.

tification in finding applications to that world ? Also in this issue the mathematical community is divided.

G.H. Hardy wrote : " I have never done anything " useful " (...) Judged by
all practical standards, the value of my mathematical life is nil (...) I have just
one chance of escaping a verdict of complete triviality, that I may be judged to
have created something worth creating. And that I have created something is
undeniable : the question is about its value "[22].

Hardy's standpoint expresses a widespread attitude in the 20th century mathematics that the highest target of mathematics is a kind of art work. Nevertheless, recently, particularly under the influence of American mathematicians a
shift towards applications is clearly noticeable and applied mathematics becomes more popular.

It happens more often than not that " pure " constructions find, sometimes
much time later, unexpected applications. Some examples : tensor calculus in
the relativity theory, operators on Hilbert spaces in quantum mechanics.

Also in this polarity we see attitudes rather complementary than opposite,
both necessary to mathematics as a whole. Nevertheless, an individual mathematician usually makes a choice.

UNITY - MULTIPLICITY

Is mathematics a single connected body, based upon an agreement concerning the subject and methodology ? Or are there more than one field which can
be justly called " mathematics " (recall the French, who say *les mathématiques* ?

Increasingly popular is the view that mathematics consists in establishing
models. The mathematical model is any (complete and consistent) set of equations corresponding to a given quantity (or to a prototype of the model) of
physical, biological, social or psychological origin, or even to some other
mathematical models.

This view seems to be supported by another one, according to which contemporary mathematics can be thought of as a collection of various theories
united only by methodological procedures. Is then mathematics a sort of " fed-
eration " ?

On the other hand, one may observe that in spite of that variety there are
very few phenomena of human culture which would be intersubjectively attain-
able to such a degree as mathematics is and which would be understood in the
same way by all users trained appropriately. The history of mathematics provides evidence that so far mathematics has been given in one version only.

22. Quoted after Ph. Davis, R. Hersh, *The Mathematical Experience, op. cit.,* 85 ff.

Looking for reasons explaining the obvious unity of mathematics one can see common mathematical principles and common basic structures. Does it suffice ? Or should we speak of " unity in multiplicity " ?

CONCLUDING REMARKS

Each polarity describes opposite attitudes towards some basic problems inherent in mathematics. The majority of active mathematicians, however, avoid reflection on the nature of their work and so their choices, which they obviously make, are usually spontaneous and unconscious. But they do make choices although it is often difficult to judge from their work what they are.

The list of polarities presented above is probably not complete and the comments provided are too short to do them full justice. Nevertheless, the exposition even in that modest form seems to be justified by an astonishing fact that mathematics does not let itself be completely attracted by any pole. It oscillates and is constantly in the field of tension between all poles. This seems to offer, at least partially, an explanation of the uncommon vivacity of modern mathematics.

THE SEARCH FOR EXPLANATIONS IN THE METHODOLOGY
OF MATHEMATICAL PHYSICS

Andrés Rivadulla

Introduction

In this paper I am concerned with two closely related topics of the philosophy of physics : the *explanation* and the *methodology* of physics. I claim both, that the methodology of mathematical physics is hypothetical-deductive, *and* that facts, empirical generalizations and other physical hypotheses sooner or later can be explained by more general laws and theories.

Contrary to any prescriptive approach to epistemology, I assume a naturalistic point of view, according to which the task of the methodology of science consists in analyzing how science actually proceeds. Nevertheless I accept, that sometimes a certain dose of normativity can be very reasonable too. In the dispute between logical positivism and critical rationalism, for example, or between relativism and scientific realism, I prefer Popper's view.

The explanation of physical entities
and the methodology of physics

To begin with, the first question we have to answer is, what kinds of physical entities need to be explained in physics. Of course, first of all the *facts*. These are *phenomena* or *events* like : planetary retrogradation in ancient astronomy, the Mars orbit, the Balmer lines, the photoelectric effect, Rutherford's deviation experience, Mercury's anomalous perihelion, the black-body radiation, the Zeeman effect, etc.

From facts to *empirical generalizations* or *empirical formulae*, there is only one step. Nevertheless philosophers of science disagree about the nature of these entities. Rudolf Carnap[1] claims for example, that " Empirical laws, ...,

---

1. R. Carnap, *Philosophical Foundations of Physics*, New York, 1966, 226-227.

are laws containing terms either directly observable by the senses or measurable by relatively simple techniques. Sometimes such laws are called empirical generalizations, as a remainder that they have been obtained by generalizing results found by observations and measurements ".

According to him the way a physicist arrives at an empirical generalization is the following[2] : " He observes certain events in nature. He notices a certain regularity. He describes this regularity by making an inductive generalization ". Boyle's law, thermal expansion, etc. are of this kind.

On the other hand Carl G. Hempel defines[3] *empirical generalizations* as those " statements of universal form that are empirically well confirmed but have no basis in theory ". Galileo's, Kepler's and Boyle's laws are some examples of them as well.

My own view on empirical generalizations, which I share with the physicist George Gamow[4], is closer to Hempel's than to Carnap's. In fact, I consider, that : " An empirical law is a physical formula that mathematically does not derive from an extant theory, but that directly follows from observations and measurements ".

Low level and *high level laws* are the next kind of physical entities that need explanation in physics. They can only be conceived as low or high level laws in terms of the relationships they maintain with each other. Thus, Max Planck's black body *radiation law* can be considered as a typical case of a high level formula, for on the basis of it many (low level) laws are explained, as we shall see later. Nevertheless it can be explained in the framework of Bose-Einstein quantum statistical mechanics.

Finally, *theories* themselves are sometimes submitted to explanation too. The explanation of theories by other ones is known as *intertheoretical reduction*. Phenomenological thermodynamics is mathematically deduced in the framework of classical statistical mechanics[5], classical mechanics is a limit case of relativity theory and of quantum physics as well, etc.

As to the different aspects of the relationship between *explanation* and the *hypothetical-deductive* character of the methodology of physics, I claim the following. To give a physical explanation of something : a novel fact, an empirical law, a low level — or even a high level — hypothesis, means to account for them in the framework of a theory. The explanation will be all the better, the more it explains, e.g., if it not only accounts for the phenomenon at issue,

2. R. Carnap, *Philosophical Foundations of Physics, op. cit.*, 228.

3. C. Hempel, *Philosophy of Natural Science*, Prentice-Hall, Inc., Engelwood Cliffs, N. J., 1966, 58.

4. G. Gamow, *Biography of Physics*, 1960, Chapter VI, footnote 1.

5. Josiah Willard Gibbs (1839-1903) undertook in his *Elementary Principles in Statistical Mechanics*, New Haven, 1902, a *rational foundation of thermodynamics* on the basis of classical statistical mechanics.

but if it also explains other phenomena, which before the explanation were considered separately ; or if it puts forward some questions which were completely unthinkable from the old point of view, etc. When this happens we say, that the theory is both *empirically and theoretically progressive*.

The methodology of mathematical physics is *hypothetical-deductive*, for only this guarantees that the explanation is given *by reference to theories*. It is hypothetical, because the explanans usually has the character of a hypothesis. It is deductive, for the explanandum has to follow deductively, *i.e.* by means of mathematical calculus, from the explanans. When this happens, the explanandum becomes a corroborating instance of the theory. Essentially this view agrees with Hempel's *deductive-nomological*[6] character of scientific explanation.

Thus the methodology of physics has nothing to do with induction. This means, that progress in physics cannot be accounted for by saying, that physicists pass from observations to theories, or, as philosophers used to say, that they get from the particular to the general ; and this amounts to claiming, that induction is not a suitable procedure for the discovery of truth. Nevertheless, many empirical formulae follow directly from observations. But contrary to Carnap's view[7] that empirical laws " are used for explaining observed facts ", I claim that *empirical generalizations* merely describe what is the case, thus falling short of giving an explanation of it. Only when physicists are able to develop theories from which these *empirical* laws *mathematically* follow, we will be right to claim that *a theoretical explanation of them has been found*[8].

Mathematics is the possibility condition of the hypothetical-deductive method of physics. Its role is by no means merely reduced to provide auxiliary tools for the formulation of laws. When empirical formulae and other hypotheses chronologically precede the theory, what very frequently happened in the history of physics, theoreticians proceed to develop, if they are lucky, theories which mathematically account for them. *Sometimes repeated failures to accomplish this task open the doors to revolutionary proposals.*

Scientific revolutions amount to *intertheoretical reductions*. The rationality of science demands that a new revolutionary theory *explains* both the failure and the success of the preceding one, which usually becomes a limiting case of the new theory. Thus, I share in this respect Karl Popper's view[9], that " a

6. C. Hempel, *Philosophy of Natural Science*, *op. cit.*, § 5.2 ; and " Aspects of Scientific Explanation ", in C. Hempel, *Aspects of Scientific Explanation and Other Essays in the Philosophy of Science*, New York, 1965.

7. R. Carnap, *Philosophical Foundations of Physics*, *op. cit.*, 227.

8. Pointing to similar circumstances Hempel claims, that " a statement of universal form, whether empirically confirmed or as yet untested, will qualify as a law if it is implied by an accepted theory (statements of this kind are referred to as theoretical laws) " (*op. cit., ibidem*).

9. K. Popper, " The Rationality of Scientific Revolutions ", in R. Harré (ed.), *Problems of Scientific Revolution*, 1975. Reprinted in K. Popper, *The Myth of the Framework. In defence of science and rationality*, Routledge, London, 1994, 21 and 22.

scientific revolution, however radical, cannot really break with tradition, since it must preserve the success of its predecessors ". Since Newtonian mechanics is explained as approximately valid, when particles move with velocities that are slow compared with the velocity of light, Popper affirms, " that Einstein's theory can be compared point by point with Newton's and that it preserves Newton's theory as an approximation ". In general terms Popper[10] points out, that " Progress in science, although revolutionary rather than merely cumulative, is in a certain sense always conservative : a new theory, however revolutionary, must always be able to explain fully the success of its predecessor. In all those cases in which its predecessor was successful, it must yield results at least as good as those of its predecessor and, if possible, better results. Thus in these cases the predecessor theory must appear as a good approximation to the new theory, while there should be, preferably, other cases where the new theory yields different and better results than the old theory ".

This view is not recent in Popper's theory of science. Already in his *Logic of Scientific Discovery*[11] he maintained this approach to scientific progress.

A CASE STUDY : THE EMPIRICALLY AND THEORETICALLY PROGRESSIVE CHARACTER OF PLANCK'S *RADIATION LAW*

In order to give an example of the hypothetical-deductive character of mathematical physics, I refer to Max Planck's black-body radiation law, which originally was derived in the framework of classical statistical mechanics. Letting S_N denote the total entropy of a system of N oscillators of frequency n and average energy U, Planck assumes that the total energy $U_N = NU$ of the system consisting of a whole number P of energy elements E is $U_N = PE$. Since the number of different forms of the distribution of P energy elements E among N oscillators is $\Omega = [(N+P-1)!]/[(N-1)!P!]$, then combining the principle of statistical thermodynamics : $S_N = kln\Omega$ with the facts that :

1) $P = NU/E$,

2) the entropy per oscillator is $S = S_N/N$, and

3) $\frac{\partial S}{\partial U} = \frac{1}{T}$, Planck obtained the mathematical expression for the *average energy U of oscillators of frequency* v.

Now, he wanted this expression to be compatible both with formula of

$$E(v) = v^3 f\left(\frac{v}{T}\right)$$

the energy density of a black body, thermodynamically given by Wien in 1894,

and with equation $E(v) = \frac{8\pi v^2}{c^3} U$, formerly obtained by Planck himself.

---

10. K. Popper, " The Rationality of Scientific Revolutions ", *op. cit.*, 12.

11. K. Popper, *The Logic of Scientific Discovery*, London, 1959. First German edition, *Logik der Forschung*, Wien, 1935, Chapter X, § 79.

Substituting $E = h\nu$ for the value of E, he successfully derived[12] the formula :

$$E(\nu) = \frac{8\pi\nu^2}{c^3}\frac{h\nu}{e^{\frac{h\nu}{kT}}-1}$$ for the radiation energy density of a black body.

Mathematical deduction of Stefan's Law

According to Planck's formula the total energy density is given by :

$$E = \int_0^\infty E(\nu)d\nu.$$

Letting $\frac{h\nu}{kT} = x$, then :
$$E = \left(\frac{8\pi h k^4}{c^3 h^4}\int_0^\infty \frac{x^3 dx}{e^x-1}\right)T^4.$$

Now since the integral value is $\frac{\pi^4}{15}$, it finally follows that $E\alpha T^4$ *i.e., Stefan's law* of the black body radiation.

In fact, in 1865 the Irish physicist John Tyndall (1820-1893) had published his experiments, according to which the radiation of a platinum wire at 1473 K was 11.7 times larger than the corresponding radiation at 798 K. In 1879 the Austrian physicist Joseph Stefan (1835-1893) found that 11.7 is approxima tively $\left(\frac{1473}{798}\right)^4$, whereof he obtained the *empirical formula* $E\alpha T^4$.

Mathematical deduction of Wien's displacement law

Max Planck's radiation law as a function of the wave length is given by :

$$E(\lambda) = \frac{8\pi hc}{\lambda^5}\frac{1}{e^{(hc/\lambda kT)}-1}.$$

If we now take $\frac{hc}{\lambda kT} = x$, then solving the equation corresponding to

$$dE/dx = 0,$$

we obtain that λT is a constant value, that we can call b.

Now, $\lambda T = b$ is precisely Wilhelm Wien's *displacement law*, discovered by him in 1893, according to which, when the temperature of a body increases, the maximum of its energy distribution moves to the shorter wave lengths.

Mathematical deduction of Rayleigh-Jeans' Law *and the* modus tollens *argument*

Finally, taking Planck's radiation law again, we can pass from quantum physics to classical physics by simply letting $h\longrightarrow 0$:

$$\lim_{h \to 0} E(\nu) = \frac{8\pi\nu^2}{c^3}kT.$$

12. As to the derivation of Planck's radiation law, see T. Kuhn, *Black-Body Theory and the Quantum Discontinuity 1894-1912*, Oxford, 1978, and M. Jammer, *The Conceptual Development of Quantum Mechanics*, 1989.

Now, this formula, known as *Rayleigh-Jeans' Law*, had already been discovered by Lord Rayleigh in 1900. It agreed very well with the experiments for large wave lengths. The *Rayleigh-Jeans' law* also reappears *mathematically* as a limiting case of Planck's radiation law.

Thus Max Planck's radiation law is empirically progressive, for, since it allows the mathematical deduction of several (different) physical laws — which previously were taken separately — about the same phenomenon, it successfully explains them.

Nevertheless, Rayleigh-Jeans' formula was very problematic. According to it, the total energy density : $E = \frac{8\pi kT}{c^3}\int_0^\infty v^2 dv = \infty$, clearly diverges.

But for high frequencies this is *completely contradicted* by experimental data, as the empirically obtained curves by the German physicists Otto Lummer (1860-1925) and Ernst Pringsheim (1859-1917)[13] for the radiation of black bodies have already made clear. A circumstance that is known as the *ultraviolet catastrophe*[14], since the Viennese physicist Paul Ehrenfest (1880-1933) used this definite description to denote it.

The situation created is captured by the following *modus tollens* logical argument

- Not accepting Planck's hypothesis $E = hv$ amounts to admitting the truth of Rayleigh-Jeans' formula.
- But this formula is experimentally falsified.

Therefore, Max Planck's hypothesis of the quantization of energy cannot be questioned any longer.

This offers an excellent example of how refutations sometimes open the doors of scientific revolutions. In fact, besides being empirically progressive, Max Planck's radiation law is theoretically progressive too, since it puts forward an idea — the quantification of energy —, that from the *classical* point of view was completely unthinkable.

SOME CONCLUSIONS RELATIVE TO THE RELATIONSHIP *THEORY-EXPERIENCE*

Let me finish by drawing some conclusions about the topics dealt with in these pages :

I.   The assimilation of empirical formulae and other kinds of hypotheses by new theories makes each former discovery " trifling ".

13. O. Lummer, E. Pringsheim, " Die Verteilung der Energie im Spektrum des schwarzen Körpers ", *Verhandlungen der deutschen physikalischen Gesellschaft*, 1 (1899), 23-41.

14. This expression appears in Section 4 of P. Ehrenfest, " Welche Züge der Lichtquantenhypothese spielen in der Theorie der Wärmestrahlung eine wesentliche Rolle ", *Annalen der Physik*, 36 (1911), 91-118. Reprinted in P. Ehrenfest, *Collected Scientific Papers*, in Martin J. Klein (ed.), Amsterdam, 1959, 185-212.

II.   Although " empirically induced " formulae may *chronologically pre-cede* the theory, their subsequent incorporation into the theory reveals the *logical irrelevance of induction* in the methodology of physics.

III.   Empirical formulae play nevertheless an important heuristic role in the development of mathematical physics, as they motivate the search for suitable theoretical explanations. Carnap himself expresses this idea as follows : " [Empirical laws] may even have motivated the formulation of the theoretical law "[15].

IV.   Besides the existence of empirical generalizations, the methods of mathematical physics help to deducing propositions about the world — among them the earlier empirical formulae — which motivate this time the design of suitable experiments in order to achieve the empirical checking of the theory itself.

V.   The relation between theory and experience is complex : sometimes the experience urges the search for explanatory theories, sometimes the theory urgently requires the design of experiments in order to be submitted itself to empirical testing.

15. R. Carnap, *Philosophical Foundation of Physics, op. cit.*, 230.

La substance évanescente de la physique

Yves Gingras

Une caractéristique frappante, mais peu souvent constatée, de l'évolution de la physique depuis la fin du XVIIe siècle, est sans contredit l'abandon progressif des concepts substantialistes comme fondement ultime de l'explication des phénomènes (tourbillons cartésiens, fluides électriques, calorique, éther, le concept de masse aussi subissant des transformations qui l'ont éloigné de l'idée initiale de " quantité de matière "). Même si l'on admettait avec John Heilbron[1] que ces fluides n'avaient qu'un caractère heuristique — ce qui n'est pas si simple — on ne pourrait nier que leur rôle a longtemps été important car ils ont fait l'objet de nombreuses recherches de la part des physiciens (en particulier dans le cas des tourbillons, de l'éther et même du calorique). Or aucune histoire de la physique ne prend pour objet et ne compare directement la transformation et la disparition de ces substances dans la longue durée.

L'idée de désubstantialisation a bien sûr été mise de l'avant il y a longtemps par les philosophes Ernst Cassirer[2] et Gaston Bachelard[3]. Bien que Bachelard ait donné plusieurs exemples de l'évolution des concepts physiques, ses analyses demeurent schématiques et ne peuvent satisfaire complètement un historien des sciences. De plus, ces auteurs n'ont pas tiré toutes les conséquences de leur point de vue. Ainsi, Cassirer insiste sur le passage d'une conception substantialiste à une conception fonctionnaliste mais il n'analyse pas le " mécanisme " qui le rend possible, comme si la mathématisation allait de soi.

---

1. J.L. Heilbron, " Weighing Imponderables and Other Quantitative Science Around 1800 ", *Studies in the Physical and Biological Sciences*, Supplement to vol. 24, Part 1 (1993).

2. E. Cassirer, *Substance and function and Einstein's Theory of Relativity*, Authorized translation by W.C. and M.C. Swabey, New York, 1953 [This book is an unabriged reprint of the volume as originally published in (Chicago) 1923].

3. G. Bachelard, *La philosophie du non : essai d'une philosophie du nouvel esprit scientifique*, Paris, 1940 ; *Le rationalisme appliqué*, Paris, 1949 ; *L'activité rationaliste de la physique contemporaine*, Paris, 1951 ; *Le matérialisme rationnel*, Paris, 1953.

Bien sûr, il existe une littérature assez abondante sur chacune de ces subs-tances (Aiton pour les tourbillons de Descartes[4], Fox pour le calorique[5] et Schaffner[6] et Cantor et Hodge[7] sur l'éther, pour ne citer que les principaux) mais rien ne suggère l'existence de relations entre ces différents secteurs de la physique, comme si ces développements étaient indépendants les uns des autres.

En fait, cette situation peut s'expliquer par le fait que, depuis une vingtaine d'années, la tendance est nettement à la " micro-histoire " des sciences et à l'analyse de cas bien délimités dans le temps et dans l'espace — autant l'espace géographique que celui des disciplines. Cette approche a bien sûr des vertus en ce qu'elle nous permet d'approfondir notre connaissance d'épisodes importants de l'histoire mais elle a comme effet pervers de faire perdre de vue les tendances à long terme de l'activité scientifique ; les vections disait Jean Piaget. Or, c'est justement dans la longue durée que notre thèse prend son sens et permet de relier entre eux des processus historiques apparemment indépen-dants.

La thèse que nous voulons défendre est qu'en prenant pour objet la relation entre mathématisation et désubstantialisation, nous pouvons voir d'un oeil nou-veau un certain nombre de problèmes classiques de l'histoire de la physique. Ainsi, dans son ouvrage *The Mechanization of the World Picture,* Dijksterhuis perçoit bien l'importance de la mathématisation de la physique mais demeure enfermé dans l'idée de " mécanisation " et rate ainsi l'impact réel de Newton qui réside justement dans l'abandon de toute tentative de " mécanisation " complète de l'univers physique. De notre point de vue, Newton offre plutôt une " dé-mécanisation de la vision du monde ", soit tout à fait le contraire de ce que suggère le titre de l'ouvrage classique de Dijksterhuis. Richard Westfall, dans son livre *The Construction of Modern Science : Mechanisms and Mecha-nics,* insiste lui aussi sur l'importance de la mathématisation, mais suggère que Newton fait la synthèse des points de vue mathématique de Galilée et mécani-que de Descartes, ce qui nous semble erroné.

Nous croyons au contraire que le débat entre Cartésiens et Newtoniens met en évidence l'incommensurabilité de leur point de vue et que le fait que la phy-sique de Newton ait fini par triompher marque une coupure historique et épis-témologique dans l'histoire de la physique dont on ne réalise pas complète-ment l'ampleur. En effet, on insiste plus souvent sur Galilée pour " dater " l'émergence de la physique moderne, alors que nous croyons plutôt que la cou-pure décisive ne survient qu'avec Newton. Ce n'est d'ailleurs qu'à la lumière

---

    4. E.J. Aiton, *The Vortex Theory of Planetary Motions*, London, 1972.

    5. R. Fox, *The Caloric Theory of Gases from Lavoisier to Regnault*, Oxford, 1971.

    6. K.F. Schaffner, *Nineteenth-Century Aether Theories*, Oxford, 1972.

    7. G.N. Cantor, M.J.S. Hodge (eds), *Conceptions of Ether : Studies in the History of Ether Theories 1740-1900*, Cambridge, 1981.

de cette opposition radicale entre les conceptions newtonienne et cartésienne de la physique que l'on peut comprendre le sort réservé à Georges-Louis Le Sage (1724-1803) et à sa tentative de synthèse de ces points de vue[8]. Notre approche permet aussi de mieux faire ressortir des continuités entre des figures comme Descartes, Le Sage et William Thomson (1824-1907). En effet, replacés dans le mouvement à long terme de désubstantialisation et de mathématisation de la matière, les travaux de Thomson sur l'éther et l'atome tourbillonnaire sont analogues à ceux des partisans des tourbillons de Descartes. L'intérêt de Thomson pour Le Sage s'explique ainsi par sa recherche d'une explication mécanique de la gravitation et n'est pas seulement une curiosité historique comme le point de vue rétrospectif le donne à penser. La difficulté d'admettre la loi gravitationnelle sans fournir d'explication est telle qu'encore en 1824, Ampère pouvait écrire que " l'attraction (newtonienne) [est] le résultat des mouvements du fluide qui remplit tout l'espace "[9], alors qu'à l'époque ce genre d'explication " cartésienne " était pourtant abandonné depuis longtemps par les physiciens.

L'incompatibilité des conceptions " chosistes ", comme dit Bachelard, et des conceptualisations mathématiques des phénomènes, se traduit chez les physiciens par une crise de l'explication qui resurgit de façon récurrente. De la réaction du *Journal des savants* à la publication des *Principia* de Newton, — selon laquelle il ne s'agissait pas de physique mais de géométrie[10] — à celle de Rutherford pour qui la mécanique quantique fait disparaître toute idée physique[11] jusqu'aux débats sur la nature " non-visualisable " de cette nouvelle physique[12], il y a un trait commun qui est la mathématisation croissante et l'abandon des modèles explicatifs fondés sur des substances. C'est d'ailleurs dans ce contexte qu'il faut replacer la critique récente faite par David Bohm[13] à la physique contemporaine d'être devenue trop " mathématique " au détriment de la partie proprement " physique ". Pour prendre un exemple encore plus récent, les différences conceptuelles entre l'électrodynamique quantique et l'électrodynamique stochastique proviennent de leur conception opposée de la nature du champ de photons virtuels qui constitue le vide. Alors que la première théorie n'y voit qu'une technique heuristique de calcul, la seconde donne à ce champ une existence réelle qui influe donc sur les forces qui s'exercent

8. S. Aronson, " The Gravitational Theory of Georges-Louis Le Sage ", in D.E. Gershenson, D.A. Greenberg (eds), *The Natural Philosopher*, vol. 3, New York, Toronto, London, 1964, 53-74.

9. Société des Amis d'André-Marie Ampère, *Correspondance du Grand Ampère*, L de Launay (publiée par), Paris, 1936-1943 (3 vols).

10. I.B. Cohen, *The Newtonian Revolution : with illustrations of the transformation of scientific ideas*, Cambridge, 1980.

11. Y. Gingras, " La physique à McGill entre 1920 et 1940 : la réception de la mécanique quantique par une communauté scientifique périphérique ", HSTC Bulletin, 5, n° 1 (1981), 15-39.

12. D. Serwer, " *Unmechanischer Zwang*, Pauli, Heisenberg, and The Rejection of the Mechanical Atom, 1923-1925 ", *Hist. Stud. Phys. Sci.*, 8 (1977), 189-256.

13. D. Bohm, D.F. Peat, *Science, Order, and Creativity*, New York, 1987.

sur les particules[14]. Sans peut-être le savoir, ces auteurs ravivent ainsi la conception électromagnétique de la nature proposée au début du siècle par Lorentz.

Notre programme de recherche vise donc à montrer qu'il existe une relation étroite entre la " mathématisation " de la physique et la " désubstantialisation " de la matière. Il s'agira de montrer que cette abstraction de plus en plus grande trouve son principe dans la manipulation des symboles mathématiques qui acquièrent une autonomie par rapport aux concepts qu'ils sont censés représenter, et en viennent à n'être définis qu'en relation avec d'autres symboles. Ce faisant, nous donnerons un contenu précis à un thème qui a souvent été abordé de façon générale mais qui n'a jamais été étudié de façon précise : le processus de mathématisation des concepts physiques.

L'idée que l'" essence " de la physique réside dans sa mathématisation croissante est bien sûr ancienne. Elle est au cœur des travaux classiques de Alexandre Koyré et de Eduard Jan Dijksterhuis sur la Révolution scientifique du XVIIᵉ siècle. Elle est aussi reprise par la plupart des historiens de la physique[15]. Cependant, il faut bien voir que ces travaux ne posent pas la question des conséquences de cette mathématisation ni des résistances auxquelles elle a donné lieu. Ils n'analysent pas non plus le fait que cette mathématisation entraîne une redéfinition complète de la physique comme discipline visant à expliquer le monde naturel. A l'exception de Norton Wise et de Enrico Bellone, très peu d'historiens ont étudié le rôle *constructeur* des mathématiques dans la définition des concepts physiques[16]. Enfin, à notre connaissance, aucun historien n'a suggéré de liens *précis* entre la mathématisation et la disparition graduelle des conceptions substantielles des tourbillons, du calorique, de l'éther et de la masse. Dans son ouvrage récent, Heilbron s'intéresse bien aux substances impondérables au tournant des années 1800 mais n'aborde pas directement les questions qui nous préoccupent ici[17].

Contrairement aux apparences donc, le processus de mathématisation de la physique n'a pas vraiment fait l'objet d'une analyse soutenue. Dans la tradition de Koyré, les travaux des historiens se sont surtout concentrés sur les influences philosophiques sur le courant de mathématisation (Pythagore, Platon, Archimède) sans vraiment aborder la question des conséquences de cette mathématisation sur la transformation de l'objet même de la physique. Prenant

14. B. Haisch, *et al.*, " Inertia as a zero-point field Lorentz force ", *Physical review A*, 49, n° 2 (February, 1994), 678-694.

15. R. Westfall, *The Construction of Modern Science : Mechanisms and Mechanics*, New York, 1971 ; T. Frängsmyr, *et al.* (eds), *The Quantifying Spirit in the 18ᵗʰ Century*, Berkeley, 1990.

16. E. Bellone, *A World on Paper. Studies on the Second Scientific Revolution*, Cambridge, 1982 ; N. Wise, " William Thomson's Mathematical Route to Energy Conservation : A Case Study of the Role of Mathematics in Concept Formation ", *Hist. Stud. Phys. Sci.*, 10 (1979), 49-83.

17. J.L. Heilbron, " Weighing Imponderables and Other Quantitative Science Around 1800 ", *op. cit.*

en quelque sorte pour acquis que la physique est mathématique, ils ont accordé peu d'attention aux résistances qu'elle a engendrées, celles-ci étant vues comme un combat d'arrière-garde. Or, les débats autour de la nature des *Principia* de Newton (traité de physique ou de géométrie ?) et du caractère occulte ou non de la gravitation, peuvent être interprétés comme l'expression d'une transformation de la définition même du terme " explication ". Pour une physique mécaniste, nécessairement substantialiste, expliquer c'est faire référence à des entités sous-jacentes aux phénomènes dont le comportement rend compte de ces derniers par des forces de contact (impulsions ou collisions). Pour une physique mathématisée comme celle de Newton, expliquer c'est plutôt invoquer les équations qui rendent compte du mouvement à partir d'un nombre limité d'axiomes. Le débat sur les causes occultes est en somme un malentendu sur le sens des mots.

En somme, ce que l'histoire de la mathématisation de la physique fait ressortir c'est le fait curieux que les mathématiques agissent en quelque sorte comme un acide sur des substances qui étaient censées rendre raison des phénomènes, comme si les raisons mathématiques s'opposaient aux raisons physiques, comme le suggérait déjà Aristote lorsqu'il écrivait dans la *Métaphysique* que " (...) L'Astronomie, en effet, a pour objet une substance, sensible il est vrai, mais éternelle, tandis que les autres sciences mathématiques ne traitent d'aucune substance, par exemple l'Arithmétique et la Géométrie. " contrairement à la physique dont l'objet est la substance même des choses[18].

LES MATHÉMATIQUES AUX COMMANDES

Mais quel est le " mécanisme " — si l'on peut utiliser ce terme ! — qui expliquerait cette désubstantialisation des concepts physiques ? Nous croyons que ce processus s'explique par l'autonomisation au sein du formalisme des symboles mathématiques représentant les variables physiques. Pour voir concrètement comment cela se passe, prenons un exemple simple : celui de la masse. Pour Newton, la masse est définie comme étant " la quantité de matière ". Comme quantité substantielle, elle est bien sûr indépendante de la vitesse, le mouvement seul ne pouvant altérer la " quantité de matière " présente en un point de l'espace. La constance de la masse est donc un corollaire de la définition intuitive et substantielle initiale. Cependant, une fois introduit dans l'équation $F = ma$, le concept de masse acquiert une certaine autonomie par rapport à la définition physique initiale en ce sens que le symbole peut être manipulé dans une suite de transformations mathématiques. Par exemple, si l'on soumet cette équation à une transformation galiléenne, elle demeure invariante. Cependant, si on la soumet à une transformation de Lorentz, les choses

18. Aristote, *Métaphysique*, XII, 1073b, 5-10.

se compliquent et suggèrent, comme on sait, que la masse varie en fonction de
la vitesse :

$$m = \frac{m_0}{\sqrt{1 - \frac{v^2}{c^2}}}$$

Si l'on s'en tenait à une lecture substantialiste, on pourrait dire que c'est là
une illusion formelle et que la masse reste en fait constante, car la quantité de
matière n'a pas changé. Si l'on adopte la définition de la masse comme inertie,
on dira plutôt que l'inertie augmente avec la vitesse, mais cette interprétation
aussi a été contestée, car l'inertie est reliée, dans la définition classique, à la
quantité de matière. Quoi qu'il en soit, cet exemple fait bien ressortir le fait
que ce sont ici les manipulations mathématiques qui génèrent la réinterpréta-
tion du concept de masse et non la réflexion sur sa nature physique.

La relation de dépendance entre la masse et la vitesse ne fait d'ailleurs pas
sa première apparition dans le cadre de la physique einsteinienne mais bien
dans celui du programme électromagnétique qui vise à rendre compte complètement de la masse en termes de champs électromagnétiques. On pourrait voir
là aussi comment les mathématiques agissent comme acide de la substance, ici
celle de masse comme quantité de matière[19]. Bien sûr les résultats numériques
diffèrent de ceux qui seront obtenus par Einstein en suivant une voie cinématique et non plus dynamique, mais le point essentiel ici est de noter que ce sont
bien les manipulations mathématiques qui rendent possibles la transformation
du concept de masse.

La puissance créative des mathématiques est encore plus frappante dans
l'histoire de la dualité onde-corpuscule. En effet, non seulement le concept de
photon ne devient-il pour ainsi dire visible que si l'on formule l'équation de
Wien en termes d'entropie de l'énergie, comme le fait Einstein en 1905, mais
la dualité onde- corpuscule ne devient elle-même pensable que lorsqu'il réussit
en 1909 à transformer l'équation de Planck dans le langage des fluctuations
d'énergie. C'est alors qu'il peut littéralement voir sur le papier que la lumière
est bien *la somme* de deux termes dont l'un représente la solution correspondant au modèle corpusculaire, l'autre celle correspondant au modèle
ondulatoire :

$$\langle E \rangle = h\nu E + \frac{c^3}{8\pi\nu^2}E^2$$

La dualité n'est en fait que l'interprétation physique de cette somme arithmétique qui n'apparaît d'ailleurs que dans le cadre d'un formalisme précis, ici
celui de la mécanique statistique.

Le temps nous manque pour multiplier les exemples ou en suivre un plus
avant, mais ces indications devraient suffire pour montrer que le mécanisme
suggéré peut rendre compte de la désubstantialisation qui accompagne la ma-

---

19. Y. Gingras, " La dynamique de Leibniz : métaphysique et substantialisme ", *Philosophiques : revue de la Société de philosophie du Québec*, XXII, n° 2, (1995), 395-405.

thématisation de concepts physiques d'abord définis intuitivement. Elles devraient aussi montrer l'intérêt d'une histoire de la physique inscrite dans la longue durée et qui renoue ainsi avec une tradition philosophique ancienne mais qui n'a pas encore épuisé sa richesse interprétative aux mains des historiens des sciences.

BIBLIOGRAPHIE SOMMAIRE

S. Bochner, *The Role of Mathematics in the Rise of Science*, Princeton, 1966.

P. Brunet, *L'introduction des théories de Newton en France au XVIIIᵉ siècle*, Paris, 1931.

D.M. Clarke, *Occult Powers and Hypotheses, Cartesian Natural Philosophy under Louis XIV*, Oxford, 1989.

I.B. Cohen, *Franklin and Newton*, Philadelphia, 1956.

E.J. Dijksterhuis, *The Mechanization of the World Picture : Pythagoras to Newton*, Translated by C. Dikshoorn, Foreword by D.J. Struik, Princeton, N. J., 1986.

P.M. Harman, *Metaphysics and natural Philosophy ; The Problem of Substance in Classical Physics*, Brighton, Totowa, 1982.

J. Hendry, " The Development of Attitudes to the Wave-Particle Duality of Light and Quantum Theory, 1900-1920 ", *Annals of Science,* vol. 37 (1980), 59-79.

M. Jammer, *The Concept of Mass in classical and modern physics*, Cambridge, MA, 1961.

M.J. Klein, " Mechanical Explanation at the End of the 19th Century ", *Centaurus*, 17 (1973), 58-82.

A. Koyré, *Newtonian Studies*, Chicago, 1968.

P. Mouy, *Le développement de la physique cartésienne, 1646-1712*, Paris, 1934, [New York, 1981].

J. Piaget, *Introduction à l'épistémologie génétique*, Tome 3, Paris, 1950 (3 vols).

K.F. Schaffner, *Nineteenth-Century Aether Theories*, Oxford, 1972.

J. Seidengart, " Théorie de la connaissance et épistémologie de la physique selon Cassirer ", in J. Seidengart (éd.), *Ernst Cassirer : De Marbourg à New York. L'itinéraire philosophique* : actes du Colloque de Nanterre, des 12-14 octobre 1988, Paris, 1990, 159-176.

J. Seidengart, " Cassirer et la philosophie des sciences en France ", *Rivista di Storia della Filosofia*, 4 (1995) 753-783.

D. Serwer, " *Unmechanischer Zwang* : Pauli, Heisenberg and the Rejection of the Mechanical Atom, 1923-1925 ", *Hist. Stud. Phys. Sci.*, 8 (1977), 189-256.

L.S. Swenson Jr., *The Ethereal Aether : A History of the Michelson-Morley-Miller Aether-Drift Experiments, 1880-1930*, Austin, London, 1972.

R.A. Watson, *The Downfall of Cartesianism, 1673-1712 : A study of epistemological issues in late 17th Century Cartesianism*, The Hague, 1966 (International Archives of the History of Ideas, II).

R. Westfall, *The Construction of Modern Science : Mechanisms and Mechanics*, New York, 1971.

E.T. Whittaker, *A History of the Theories of Aether and Electricity*, New York, 1989 (2 vols).

R.S. Woolhouse, *Descartes, Spinoza, Leibniz. The Concept of Substance in 17th Century Metaphysics*, London, 1993.

De l'histoire des techniques
à la philosophie de la technologie

Jean C. Baudet[1]

Le problème du rapport entre Science et Technique est central en épistémologie, d'autant plus peut-être aujourd'hui que, sous la forme des " nouvelles technologies ", la technique est massivement présente dans les préoccupations de notre temps. Comme cela a souvent été dit, cette omniprésence de la technique s'accompagne paradoxalement d'un important déficit de la réflexion philosophique, comme si le fait technique était oublié, voire forclos[2] par le philosophe. Nous espérons apporter ici un élément de solution à cette question, en partant d'une analyse des (rares) considérations sur la nature de la technique faites par les historiens des techniques. Notre recherche visant à déterminer le statut épistémologique de la technique, elle est fortement marquée par notre prise de position philosophique, ce qui signifie que nous ne revendiquons d'aucune manière l'objectivité de l'historien ou du sociologue. La question est bien " qu'est-ce que la technique par rapport à la science ", non à une époque donnée, comme pourraient le déterminer les historiens, ni dans une société donnée, comme pourraient le préciser les sociologues, mais *en soi*, c'est-à-dire métaphysiquement.

Nous réglons la question terminologique préalable (technique-technologie) par les définitions suivantes. La technologie est l'ensemble des moyens dont disposent les ingénieurs pour résoudre les problèmes qui leur sont posés. La technique est l'ensemble des moyens dont disposent les artisans pour résoudre les problèmes qui leur sont posés. Quand cela ne donne pas lieu à ambiguïté, nous utiliserons également le terme " la technique " pour désigner l'ensemble technique-technologie. Nous résolvons donc la question de la distinction entre

1. J.C. Baudet, " Ambiguïté des relations entre science et technologie ", *Technologia,* 1 (1), (1978), 17-20.
2. C'est G. Hottois qui a utilisé le terme de forclusion, emprunté à la psychanalyse, pour caractériser l'élimination de la technique de son champ de conscience par le philosophe. Voir G. Hottois, *Le Signe et la technique*, Paris, 1984.

technique et technologie en la ramenant à la question de la différence entre
l'artisan et l'ingénieur. Celle-ci peut, en première approximation, être ramenée
à la question *historique* de savoir à quel moment est apparu l'ingénieur
" moderne ". L'on peut admettre que ce moment se situe lors de la Révolution
industrielle anglaise, vers la fin du XVIIIe siècle. Une discussion historique sur
le choix de cette date, qui pour nous n'est finalement qu'un simple repère chro-
nologique, sortirait du cadre de notre propos.

Rappelons qu'à côté de ce sens de technique " avancée ", le terme technologie a eu deux autres sens. L'on peut remonter à Christian Wolff, qui en 1728
proposait la définition suivante : *Technologia est scientia artium et operum
artis*[3], pour constater que de nombreux auteurs ont entendu par technologie
une description générale, une science donc, des différentes techniques (les arts
et métiers, les arts de l'ingénieur, et expressions semblables). A ce sens qui est
basé sur la généralité, s'oppose un deuxième sens que l'on retrouve dans les
écoles d'ingénieurs, sens basé sur la spécificité. Il s'agit ici, par technologie,
de désigner les modalités pratiques de mise en oeuvre des principes techniques. Par exemple, la " technologie de la construction " serait l'ensemble des
procédés et tours de main par lesquels sont réalisés les systèmes constructifs
(murs, toitures, ponts...) conçus et calculés par les ingénieurs. La " technologie
du soudage " serait l'ensemble des procédés et tours de main des soudeurs, etc.
Dans ces deux acceptions, la technologie est donc soit une réflexion généralisante (essentiellement une classification) basée sur l'ensemble des techniques,
soit au contraire une partie spécifique de chaque domaine technique.

Notons que ces deux sens tendent à sortir de l'usage. Pour un vaste public,
aujourd'hui, la technologie c'est bien l'ensemble des techniques actuellement
disponibles, c'est le savoir des ingénieurs, comme la science est l'ensemble
des connaissances actuellement acquises, c'est le savoir des chercheurs.

Qu'en est-il alors de la philosophie de la technologie ? Signalons d'abord,
pour n'y plus revenir, qu'un auteur français, Alfred Espinas (1844-1922), a
proposé à la fin du XIXe siècle d'utiliser le terme technologie (ou praxéologie,
parfois orthographié praxiologie) pour désigner une philosophie des techniques
dont il notait la nécessité, avec même l'idée de constituer plus globalement une
philosophie de l'action[4]. Cette idée sera notamment reprise par le philosophe
polonais Thadée Kotarbinski[5].

Notre point de départ étant la fin du XVIIIe siècle, nous pouvons retenir
comme débuts de la réflexion sur la technologie les trois dates suivantes que
nous proposent les historiens des techniques. 1771 : John Smeaton (ingénieur

---

3. Cité par J. Sebestik, " The rise of the technological science ", *History and Technology*, 1
(1983), 25-44, 27.

4. A. Espinas, " Les origines de la technologie ", *Revue philosophique de France et de l'Etranger*, 30 (1890), 113-135 ; 295-314.

5. Th. Kotarbinski, " Les problèmes de la praxiologie ou théorie générale de l'activité efficace ", *Revue philosophique de la France et de l'Etranger*, 154 (4), (1964), 453-472.

anglais, 1724-1792) fonde la Society of Civil Engineers. 1777 : Johann Beckmann (professeur à l'Université de Göttingen, 1739-1811) publie une *Anleitung zur Technologie*. 1794 : fondation à Paris de l'Ecole Polytechnique, sous l'impulsion du géomètre français Gaspard Monge.

L'on dispose donc, pour réfléchir sur la technologie en train de naître, d'une association de professionnels en Grande-Bretagne, d'un livre en Allemagne, d'une école en France.

UNE APPROCHE ENCYCLOPÉDIQUE

Jan Sebestik[6] a montré comment, à partir de la parution de l'ouvrage de Beckmann[7], une " science des techniques " se constitue, sous le nom de technologie[8], à l'articulation des XVIIIe et XIXe siècles. Il s'agit d'une science-description, avec déjà chez certains auteurs un début de réflexion que l'on peut qualifier de protophilosophique.

Certes, l'ouvrage de Beckmann est encore exclusivement descriptif, dans la tradition des encyclopédies du XVIIIe siècle : les différents articles sont simplement juxtaposés, d'ailleurs par ordre alphabétique. L'objectif de l'auteur est de décrire les techniques, l'une après l'autre, et apparemment sans se soucier de ce qu'elles ont en commun. Mais en 1806 il publie un *Entwurf der algemeinen Technologie* dont l'ambition est clairement plus fondamentale[9].

De l'oeuvre de l'Allemand, l'on peut rapprocher, en France, le livre de Gérard-Joseph Christian (directeur du Conservatoire Royal des Arts et Métiers) : *Plan de Technonomie*. Le terme technonomie, concurrent de technologie, ne survivra pas à son inventeur. Christian est déjà à la recherche de ce qui caractérise et fonde les diverses techniques. Il ne se satisfait plus des ouvrages purement descriptifs de son temps : " Ce n'est pas toutefois qu'on puisse regarder ces traités de technologie comme autre chose qu'un arrangement, une simple classification systématique des procédés des arts. On y chercherait en vain des faits généraux, des déductions théoriques et fécondes, en un mot, une doctrine qui les domine et les embrasse tous "[10]. Il trouvera d'ailleurs cet élément général, qui réduit les diverses techniques (les arts) à n'être que des variations d'un même phénomène. " Toute opération manufacturière consiste uniquement dans l'emploi d'une seule chose : la force "[11]. Traduisons : la

---

6. J. Sebestik, " The rise of the technological science ", *op. cit.*

7. J. Beckmann, *Anleitung zur Technologie oder zur Kenntnis der Handwerke, Fabriken und Manufacturen*, Göttingen, 1777.

8. " Technologie " en France et en Allemagne. *Technology* aux Etats-Unis d'Amérique. Le terme *technology* semble avoir été peu usité à cette époque en Angleterre.

9. J. Beckmann, *Entwurf der algemeinen Technologie*, Göttingen, 1806.

10. G-J. Christian, *Vues sur le système général des opérations industrielles ou Plan de technonomie*, Paris, 1819, 35.

11. *Idem*, 93.

force = l'énergie. Aux Etats-Unis, dans un souci du même ordre, paraissent les *Elements of Technology* de Jacob Bigelow[12], qui enseigne à Harvard.

L'on peut également rattacher à cet effort encyclopédique accompagné d'un début de réflexion (notamment sur les implications sociales de la technologie), en Grande-Bretagne, l'ouvrage de Charles Babbage (le célèbre " inventeur " de l'ordinateur) *On the Economy of Machinery and Manufactures*[13] et même *The philosophy of manufactures* d'Andrew Ure[14]. Le terme *philosophy* ne doit évidemment pas être traduit par philosophie, mais plutôt par " description systématique ", conformément à sa signification de l'époque en Grande-Bretagne. Malgré son titre très général, l'ouvrage d'Ure se limite en fait au secteur textile, en étudiant les manufactures traitant le coton, la laine, le lin et la soie.

UNE APPROCHE MATHÉMATISANTE

A côté de ce courant encyclopédique qui aurait pu être le lieu de naissance d'une authentique philosophie de la technologie, il importe d'étudier l'effort de clarification mené à l'Ecole Polytechnique. Le travail initiateur est ici l'ouvrage de Lazare Carnot, *Essai sur les machines en général*[15]. Viennent ensuite le *Traité élémentaire des machines* de Jean Hachette[16], *Du calcul de l'effet des machines* de Gaspard-Gustave de Coriolis[17], et surtout le *Cours de mécanique appliquée aux machines* de Jean-Victor Poncelet[18]. Comme l'a bien montré Konstantin Chatzis[19], un des grands débats pédagogiques au sein de l'Ecole Polytechnique pendant son premier demi-siècle d'existence aura pour objet la manière d'articuler le cours de mécanique rationnelle (purement mathématique) et le cours de machines. Chatzis a montré comment la réflexion sur ce problème a conduit à des exposés qui vont aboutir à développer une théorie des mécanismes, qui constituera la base conceptuelle de la technologie, du moins dans le domaine mécanique. Alors qu'au départ les machines sont analysées à partir de l'idée d'équilibre (primat de la statique), petit à petit les enseignants de l'Ecole se rendent compte qu'il est plus fructueux de partir, pour l'étude des machines (essentiellement la machine à vapeur et les machines hydrauliques), du mouvement. Ainsi la cinématique se constitue-t-elle, grâce notamment à l'oeuvre de Jean-Baptiste Bélanger[20], alors même que se constituait par

---

12. J. Bigelow, *Elements of Technology*, Boston, 1829.

13. Ch. Babbage, *On the Economy of Machinery and Manufactures*, Londres, 1832.

14. Λ. Ure, *The philosophy of manufactures or An Exposition of the Scientific, Moral, and Commercial Economy of the Factory System of Great Britain*, Londres, 1835.

15. L. Carnot, *Essai sur les machines en général*, 1782.

16. J. Hachette, *Traité élémentaire des machines,* 1811.

17. G.-G. de Coriolis, *Du calcul de l'effet des machines*, 1829.

18. J.-V. Poncelet, *Cours de mécanique appliquée aux machines*, 1836.

19. K. Chatzis, " Mécanique rationnelle et mécanique des machines " dans B. Belhoste *et al.* (eds), *La formation polytechnicienne 1794-1994*, Paris, 1994, 95-108.

20. J.-B. Bélanger, *Cours de mécanique*, 1847 ; J.-B. Bélanger, *Traité de cinématique*, 1864.

ailleurs la thermodynamique, avec l'oeuvre de Sadi Carnot[21], également à partir de réflexions générales sur le fonctionnement des machines.

Il y a donc, vers le milieu du XIXe siècle, deux approches concurrentes d'un début de réflexion générale sur les techniques.

A l'Ecole Polytechnique, et dans les écoles d'ingénieurs qui se réclameront du modèle polytechnique (primat de la mathématisation), la base de la technique est une théorie des mécanismes : les objets techniques sont des corps en mouvement (ou en repos, considéré comme un cas particulier de mouvement, quand v = 0). L'ingénieur allemand Franz Reuleaux (1829-1905)[22], qui fut directeur de l'Académie industrielle de Berlin, synthétisera cette approche dans des ouvrages qui seront largement diffusés.

L'autre approche, qui est encore celle des encyclopédistes, voire celle des auteurs de *Théâtres des machines*, est essentiellement descriptive et énumérative. C'est sans doute l'approche utilisée dans l'enseignement des écoles d'ingénieurs du type " arts et métiers ".

LES CLASSIFICATIONS DES SCIENCES

Un troisième courant de réflexions aurait dû, en cette première moitié du XIXe siècle, conduire à une analyse de la technique : les multiples propositions de classification des sciences. Dès 1802, Claude Henri de Saint-Simon, dans scs *Lettres d'un habitant de Genève*[23], adopte la suite " astronomie - physique - chimie - physiologie " dont on sait qu'elle formera le noyau de la classification des sciences d'Auguste Comte. Cette question d'une échelle des disciplines scientifiques classées par ordre de complexité croissante et de généralité décroissante fera l'objet de nombreux travaux au XIXe siècle mais, de Comte à Engels, en passant par André-Marie Ampère, Antoine-Augustin Cournot, et d'autres, elle ne prendra jamais en considération la technique, qui sera soit complètement écartée du champ de la science, soit considérée comme une science " appliquée ", c'est-à-dire ayant, par rapport à la science, un statut épistémologique dégradé. Avec Marx, Engels avait nettement perçu que la technique est en fait fondatrice de la science[24], il est dès lors étonnant qu'il ne prend pas davantage en compte la technique dans sa classification, qui reprend essentiellement celle de Comte, en y ajoutant la dialectique avant la mathéma-

---

21. S. Carnot, *Réflexions sur la puissance motrice du feu et sur les machines propres à développer cette puissance,* Paris, 1824.
22. Fr. Reuleaux, *Cinématique,* Paris, 1877 (traduction).
23. C.H. de Saint-Simon, *Lettres d'un habitant de Genève à ses contemporains,* 1803.
24. K. Axelos, *Marx, penseur de la technique,* Paris, 1961 ; B. Kédrov, *La classification des sciences,* Moscou, 1977-1980 (2 vols).

tique et l'histoire après la sociologie. Ni Piaget[25], ni Kédrov[26] ne situeront davantage la technique dans le champ de la science.

HEIDEGGER ET SA POSTÉRITÉ

La pensée sur la technique au XXe siècle est dominée par la figure de Martin Heidegger. Le 18 juillet 1962, il donnait une conférence devant un public d'ingénieurs et d'étudiants-ingénieurs qui constitue une méditation sur le rapport entre technique et non-technique, dont le texte a été récemment édité, puis traduit en français[27]. Ce texte permet de saisir la pensée du philosophe sur la technique, qui a — dès la parution de *Sein und Zeit*, en 1927 — influencé pratiquement tous les courants de pensée anti-techniciens à caractère pessimiste (Herbert Marcuse, Jürgen Habermas, et même Jacques Ellul). Sommairement, nous rappellerons qu'Heidegger voit dans la technique le phénomène fondamental de la modernité, mais qu'en même temps il la dénonce car conduisant à ce qu'il appelle un " oubli de l'être ". Il est particulièrement frappant de constater combien les philosophes américains contemporains qui s'occupent de technique sont fascinés par la pensée heideggerienne. Tant les collaborateurs à la revue électronique *Techné : Journal of the Society for Philosophy and Technology*, animée par Paul T. Durbin[28], que ceux qui publient dans la revue électronique *Tekhnema*, animée par Richard Beardsworth[29], développent souvent leur pensée à partir d'Heidegger, qu'il s'agisse d'ailleurs d'accepter ou de rejeter les thèses du fondateur de l'existentialisme. Notons aussi qu'en dépit de ce point de départ métaphysique, la production philosophique américaine traitant de la technique est surtout de préoccupation éthique ou politique, s'attachant rarement à chercher l'*être* de la technologie.

Après avoir rappelé le caractère instrumental de la technique[30], Heidegger découvre que la technique est un concept du savoir, pas un concept du faire, ce qui l'amène à devoir admettre un lien " insoupçonné " entre science et technique, à savoir que " c'est la science qui dérive de la technique ". La technique, dira Heidegger, " est codéterminante dans le connaître "[31].

L'antériorité existentielle de la technique sur la science sera par exemple réaffirmée par le philosophe américain Don Ihde[32], pour qui la technique est

25. J. Piaget, " Le système et la classification des sciences " dans J. Piaget (éd.), *Logique et connaissance scientifique*, Paris, 1967, 1151-1224.

26. B. Kédrov, *La classification des sciences*, op. cit.

27. M. Heidegger, *Langue de tradition et langue technique*, Bruxelles, 1990 (posthume).

28. Fondée en 1996, la revue électronique *Techné*, dont Paul T. Durbin (Université du Delaware) est le rédacteur en chef, est un organe de la Society for Philosophy and Technology, fondée en 1977.

29. La revue *Tekhnema* a été fondée par l'American University of Paris en 1993.

30. M. Heidegger, *Langue de tradition et langue technique*, op. cit., 21 et sq.

31. M. Heidegger, *Langue de tradition et langue technique*, op. cit., 25.

32. D. Ihde, *Technics and Praxis*, Dordrecht, 1979.

" une façon d'être au monde ". Un autre auteur américain, Frederick Ferré, retrouvera la continuité métaphysique qui lie la science à la technologie, en partant d'une intéressante comparaison entre le discours religieux *as value-expressing but with an ineliminable referential or metaphysical component* et le discours scientifique *as referential but with an ineliminable general context that is profoundly value-laden*[33].

LA TECHNOLOGIE RETROUVÉE

Cette codétermination du connaître par la science et par la technique nous autorise à situer, épistémologiquement (c'est-à-dire en pensant ce positionnement non comme une simple commodité de méthode, mais comme le résultat d'une profonde identité ontologique entre le dit de la science et le fait de la technique), la technologie parmi les diverses disciplines au statut scientifique reconnu. Ne considérant la technologie ni comme " introuvable "[34], ni comme située dans un ailleurs de la science, ni même comme une science dégradée par son incarnation dans des " applications " qui ne seraient que " utilitaires ", nous la plaçons nettement quelque part sur l'échelle d'Auguste Comte, qui malgré les nombreux travaux critiques qu'elle a provoqués reste utilisable. Mais où situer la technologie par rapport à la logique, la physique et la sociologie, puisque l'échelle comtienne se réduit aujourd'hui à ces trois disciplines, la mathématique étant rattachée à la logique, et les spécificités qui justifiaient, au temps de Comte, les distinctions entre physique et chimie (l'affinité) et entre chimie et biologie (la force vitale) n'étant plus pertinentes pour la science dans son état actuel ?

Ce qui, dans le système de Comte, justifie la position d'une science C immédiatement après une science B, c'est d'une part la continuité épistémique qui va de B à C, et d'autre part la reconnaissance d'un élément dans l'objet de C irréductible aux éléments constituant l'objet de B (par exemple, au début du XIXe siècle, l'on pouvait en effet considérer que la force vitale, qui distingue les êtres vivants des objets inertes, était irréductible aux lois physico-chimiques, ce qui justifiait la séquence chimie-biologie).

Nous plaçons la technologie *après* la sociologie, et nous avons donc la séquence logique-physique-sociologie-technologie. Par sociologie, nous entendons évidemment l'ensemble des " sciences de l'homme ", puisqu'en effet la conscience peut être (quelle que soit d'ailleurs la position métaphysique sous-jacente à cette considération) envisagée comme irréductible aux lois de la physique, et que c'est bien la conscience qui caractérise le phénomène humain. La technologie se situe alors à la fin de l'échelle de Comte réhabilitée, parce que l'outil en constitue l'élément irréductible. Un outil est en effet non seulement

---

33. F. Ferré, " Philosophy and Technology After Twenty Years ", *Techné*, vol. 1, 1-2 (1995).
34. J-C. Beaune, *La technologie introuvable*, Paris, 1980

plus qu'un simple objet matériel (étudiable par la physique), mais également plus qu'un simple objet culturel (étudiable par la sociologie). Il sortirait du cadre du présent article d'aller plus loin dans l'analyse ontologique de ces *étants* particuliers que sont les outils, qui entretiennent une relation forte avec le concept de communication[35].

Notre affirmation de la technologie comme science à part entière trouve sa justification d'abord dans le vécu des chercheurs et des ingénieurs, qui utilisent les mêmes schémas de pensée, qu'il s'agisse de comprendre une galaxie, les liens de parenté dans une société amazonienne, ou une centrale électronucléaire. Quant à sa position après la sociologie, elle est justifiée par le fait que le facteur humain, pour l'ingénieur, est une " donnée du problème " au même titre que la résistance des matériaux ou la sécurité des approvisionnements. Le management et le marketing sont devenus des parties intégrantes de la technologie, comme cela ressort clairement de la pratique industrielle et des programmes actuels des écoles d'ingénieurs. La technologie, placée ainsi en fin d'échelle, est une science humaine[36], une science de synthèse. Ce n'est pas une science appliquée. C'est une science achevée.

35. La nature communicationnelle de l'outil (qui est toujours un interface entre l'homme et son désir) est une autre entrée dans la reconnaissance de la conature entre technique et science, puisque celle-ci est un discours. Voir J.C. Baudet, " L'éditologie : entre communication et cognition ", *Revue générale*, 4 (1997), 45-54.

36. J-C. Baudet, " La technologie est une science humaine ", *European Journal of Engineering Education*, 18 (3), (1993), 293-299.

THE HISTORY OF SCIENCE AND THE ETHICS OF SCIENCE

Martin COUNIHAN

INTRODUCTION

It is obviously important for teachers at all levels to have clear conceptions of the aims of their work : if we do not know what we are trying to achieve, then we shall not be able to assess how well we achieve it. There are a number of purposes for which the history of science might be taught. One possible reason was given recently by Torkil Heiede : referring specifically to the question *Why Teach History of Mathematics ?* he asserted disarmingly that the history is inextricable from the subject itself :

" If you teach mathematics, you must also teach history of mathematics, for the history of a subject is part of the subject.

If you are not aware that mathematics has a history, then you have not been taught mathematics — because you have then been cheated of an indispensable part of it "[1].

As mathematics in particular, so science in general. Heiede's view is philosophically interesting because, if he is right, scientific knowledge has no culturally-independent core which could be taught as a logical and experimental system, justified in its own terms, abstracted from the historical context in which it happened to emerge on this planet. In effect, Heiede is making a statement against the philosophical stance of scientific realism and in general support of the kind of social constructivism which has been given prominence by the recent " Science Wars " debate[2].

If Heiede's approach were accepted, the aims of teaching the history of science would be reduced to the aims of teaching science itself. This convergence of aims has been highlighted with a recent claim, in a different context, that

---

1. T. Heiede, " Why Teach History of Mathematics ? ", *The Mathematical Gazette*, vol. 76, n° 475 (March 1992).

2. See *Social Text*, 46-47 (1996).

the history of science does not play the significant role that it should in education and training in science in Europe, and that the history of science should be used to help us to train better scientists[3].

However, this paper is a discussion of a somewhat different purpose for the teaching of the history of science, namely that of conveying to students something of the ethics of science. It has been written in the context of an ICSU-sponsored project on *Science, Ethics and Education* intended to generate educational materials on the ethics of science. This project has led to a book by Fullick and Ratcliffe[4] which draws on the contemporary and recent history of science to promote ethical debate in the classroom and to suggest a broadly Mertonian ethos (communal, universal, disinterested, sceptical) in harmony with ICSU's statutes concerning universality, freedom, and non-discrimination.

After considering the work of Fullick and Ratcliffe below, a classic paper by Langmuir[5] is briefly discussed. This is followed by a consideration of the tension between the ethics of science and contemporary approaches to the history, philosophy and sociology of science. This entails a review of the different disciplines that have been promoted by different people at different times as a " metadiscipline " for the analysis of knowledge itself. Finally, some tentative conclusions are drawn in section 7, although the objective of this paper is to draw attention to some neglected issues rather than to resolve them.

Some Definitions

For our present purposes, " science " is defined as the understanding of nature and of humanity's place in it. Technology, *i.e.* the manipulation and control of nature, is not being considered here. So, the " ethics of science " is not intended to include issues to do with the impact of technology. Important as they are, technological, environmental and medical ethics are beyond the scope of this paper.

" Morals " are taken to be the set of generally-accepted norms which govern the behaviour of a particular group of people. Defined in this way, morals can be regarded as a social fact, a set of rules which can be established by observing or questioning the group of people concerned. " Ethics " is the study or justification of morals. The term sometimes causes confusion because morals can be studied from various points of view and for various purposes. Ethical study

---

3. *Cf.* the Conference on the *History of Science and Technology in Training in Europe, Strasbourg, June 1998*.

4. P. Fullick, M. Ratcliffe (eds), " Teaching Ethical Aspects of Science ", *Science, Ethics and Education Project* (Committee on the Teaching of Science of the International Council of Scientific Unions), Southampton, 1996.

5. F.A.J.L. James, " Faraday in the Pits, Faraday at Sea : the role of the Royal Institution in changing the practice of science and technology in nineteenth-century Britain ", *Proc. Roy. Inst.*, vol. 68 (1997), 277.

is essentially an interdisciplinary activity, drawing on the methods and insights of many disciplines — theology, philosophy, history, sociology and others.

This paper is about the use of the history of science in the context of the ethics of science. The " ethics of science " is taken to be the study of the morals which might govern the behaviour of scientists " as scientists ", including what might be the characteristics of an " ideal " scientific community. We will not be concerned here with the behaviour of other people such as the politicians who fund science, the teachers and others who communicate science to non-scientists, or the public who may or may not have proper reverence for science, although their activities certainly involve important ethical issues.

THE UNIVERSALITY OF SCIENCE

What can we say about the ethics of science ? The " universality " of science is often taken as a fundamental guiding principle. ICSU, the International Council of Scientific Unions, declares :

" As the intrinsic nature of science is universal, its success depends on co-operation, interaction and exchange, much of which goes beyond national boundaries. In order to achieve its objectives, scientists involved in ICSU activities must therefore be able to have free access to each other and to scientific data and information. It is only through such access that science can produce its fruits and international scientific co-operation can flourish ".

The universality principle is very ancient and asserts the inter-ethnic and inter-cultural character of science. A " science " which is esoteric or narrowly ethnocentric is not, in the final analysis, really science at all. The universality principle can be confusing because it is not the same as the idea that a true scientific law should apply across the whole universe. Nor is it the same as the theory of epistemological convergence, *i.e.* that there can only be one ultimate science, which follows if we accept that science is an objective representation of the universe and not merely a social construction.

CASE STUDIES IN THE HISTORY OF SCIENCE

In the work of Fullick and Ratcliffe[6] a number of case studies are given through which, it is suggested, students can explore issues in the ethics of science. The case studies are all from recent or contemporary history and include the use of methyl bromide as a pesticide, whether ethanol should be the fuel of the future, our personal use of energy, the consequences of the use of cars, and the counselling of carriers of Tay-Sachs disease. There is a very strong emphasis here on technology and medicine : in fact the authors would evidently agree with the view that " science and technology are now so inextrica-

6. P. Fullick, M. Ratcliffe (eds), " Teaching Ethical Aspects of Science ", *op. cit.*

bly linked that it is impossible to try to separate them "[7], which goes back to Francis Bacon's dictum that " knowledge and human power are synonymous " and which is now once again fashionable among philosophers, historians and sociologists of science.

For many purposes — for example, the study of the social context of science — it is indeed generally true that science cannot be disentangled from technology. However, I would argue that it is precisely when one considers the *ethics* of science that the distinction becomes relevant and in fact crucial. The understanding of nature and the controlling of nature are different motivations supported usually by different ideologies. Yeats put it well :

> " The intellect of man is forced to choose
>
> Perfection of the life, or of the work,
>
> and if it take the second must refuse
>
> A heavenly mansion, raging in the dark "[8].

Notwithstanding their technological emphasis, Fullick and Ratcliffe include some case studies on " pure " science, looking at the Lysenko and cold fusion affairs and at the work of Crick and Watson on DNA. Their stance is to use these case studies to support the four " Norms of Science " associated with Merton[9], *i.e.* communalism, universalism, disinterestedness, and organised scepticism.

Communalism (science is for all humanity) and universality (the integrity of the scientific community) were strongly endorsed by ICSU in the quotation given earlier above. The principle of disinterestedness claims that science is in a certain sense neutral. " Organised scepticism " amounts to an insistence on criticality and unbending rationality in science. It is interesting that, while earlier generations of scientists might have taken these principles for granted, some or all of the four would today be seriously questioned by many scholars — for example, the " special " place of rationality in science would be questioned by those who challenge the sense of any " rational reconstruction " of the history of science.

Another example of how the recent history of science has been used to inform the ethics of science is Langmuir's famous paper on *Pathological Science*[10]. Langmuir's case studies include the Davis-Barnes effect, Blondlot's N-rays, the mitogenetic rays studied by Gurwitsch and others, the Allison effect, and Rhine's studies of extrasensory perception. His objective was to show how the scientist — and whole groups of scientists — can all too easily be led into

7. F.A.J.L. James, " Faraday in the Pits, Faraday at Sea : the role of the Royal Institution in changing the practice of science and technology in nineteenth-century Britain ", *op. cit.*

8. From W.B. Yeats, *The Choice*, 1932.

9. P. Fullick, M. Ratcliffe (eds), " Teaching Ethical Aspects of Science ", *op. cit.*, 92.

10. I. Langmuir, " Pathological Science ", in R.N. Hall (ed.), *General Electric Company report 68-C-035*, New York (April 1968).

error by loose methodology, a lack of objectivity, the failure of a sub-community of scientists to take proper notice of the scientific community as a whole, and the allure of personal reputation. These factors are all echoed in the Mertonian " Norms of Science " mentioned above, although Langmuir's main emphasis is on the ethic of objectivity.

It is interesting that Langmuir's use of history consists of looking not at good and successful science but at bad science. It is the charm of his paper that he looks into murky corners and tells unfamiliar stories. Who now remembers that the Allison effect led to the discovery of half-a-dozen new chemical elements, including Alabamine and Virginium ? Fullick and Ratcliffe, too, focus on bad science such as the cold fusion affair to draw out ethical lessons.

Ethicists of science such as these are much more sophisticated than to restrict their attention to successful science or to present a Whiggish description of how the great scientists discovered the truth : however, in other respects they appear to be at odds with contemporary approaches to the history of science. Today it seems naive to suggest that there is a " single " generally-accepted morality of science. Since the time of Thomas Kuhn it has been commonplace to say that science passes through paradigm-shifts which are ethical and social and well as internally scientific in character. Many if not most scholars would now claim that the history of science does not show sufficient continuity or universality for a single morality of science to be abstracted from it ; that it is futile to try to construct a culturally-neutral basis for the ethics of science ; and that the history of science, far from being useful for demonstrating an ethical basis for science, actually demonstrates the lack of any such basis.

AGAINST HISTORICISM ?

From the preceding sections it should be clear that it is problematic to use the history of science for ethical purposes, *i.e.* as a justification for particular morals in science. Contemporary science studies may have aggravated matters, and would claim that the history of science is a socially-constructed and subjective narrative, profoundly unreliable for informing the ethics of science, but the problem is not new : indeed Karl Popper's critique[11] of historicism would suggest that the history of science cannot be expected to yield any sort of universal law or inevitable trend in the development of science — although in saying so we may be going beyond Popper's intention. It would follow that the history of science, whatever its other uses, cannot be taken to justify a law of scientific progress or a particular methodology associated with such progress. That is not to say, of course, that such laws or methods do not exist : it simply says that history is not the discipline that will reveal them. It might be perfectly

11. K.R. Popper, *The Poverty of Historicism*, 2nd ed., London, 1960.

possible for philosophy, say, or theology to reveal the pattern of human progress or the " proper " method of science. But we cannot expect it to come out of the history of science !

But is this reasonable ? Pursuing the Popperian theme, we could say that historicism suffers from the same circularity as inductivism, no more and no less. The principle of induction can be justified illogically on the grounds that inductive inference has proved effective in the past. Correspondingly, if the historicist analyses the history of science with scientific rigour, then (by Popper's own methodological insights) any resulting universal law of scientific progress can be no more than a tentative not-yet-refuted hypothesis, and not therefore a reliable prediction, nor an absolute basis for scientific morals. But it does not follow that the history of science is misleading, any more than science itself is misleading.

OTHER DISCIPLINES

It is necessary to look broadly at the history of the inter-relationships and rivalries between academic disciplines in order to see what is going on here. The " Science Wars " attack by social theorists can be seen as merely the latest in a long line of attempts by different disciplines each to assert that they have a superior epistemological status and that all other disciplines are in a sense subsidiary to them.

Theology was traditionally regarded as the " Queen of the Sciences " because the subject matter of theology — God — creates and transcends everything.

The decline of theology during and since the Enlightenment left physics as the key discipline. As Bonaparte claimed, " they may say what they like : everything is organised matter ". Some theologians responded not by opposing physics on its own territory but by arguing dualistically that theology and physics deal with completely separate realms. As John Henry Newman put it in the mid-19th century, " theology and physics cannot touch each other, have no intercommunion, have no ground of difference or agreement, of jealousy or of sympathy ".

Mathematics, a hundred years ago, could make a powerful claim to be the key discipline. It seemed likely that rationality itself — on which all intellectual activity depends — could be supported by the logical foundations which mathematics was building for itself. Indeed the influence of mathematics has been extended enormously during the 20th century as the majority of academic disciplines have become quantified, but Gödel's theorems are perceived to have undermined its foundational claims.

History is another discipline to have claimed the central place in human knowledge — and, after all, everything that has ever happened in recorded human experience is the province of history. For the defects of historicism, however, we need look no further than Popper[12].

Philosophy, of course, is the discipline which has made the most insistent claims to pre-eminence. Post-modern philosophy deconstructs all disciplines except itself. In its alliance with contemporary social theory, it underlies the critique of science associated with the phrase " Science Wars "[13] and alluded to by Charles Gillispie in his Plenary Talk at the start of this conference.

Although the " Science Wars " affair has been most obviously a controversy between the new " science studies " and physics, the broader controversy has also involved theology. Social theory attempts to transcend, deconstruct and relativize theology just as it does science. Ian Ker and Alvin Plantinga are among the leading theologians today arguing against what they call " creative anti-realism ", i.e. against the post-modern sociological view that denies the reality of divinity at the same time as it denies scientific realism, and thereby treats both theology and science as mere social constructions.

As Plantinga puts it, reported by Ker[14], creative anti-realism says that " we human beings, in some deep and important way, are " ourselves " responsible for the structure and nature of the world ; it is we, fundamentally, who are the architects of the universe. "[15] The philosophers and sociologists, replacing " divine with human creativity "[16], and have promoted relativism and the loss of objective truth.

Biology is yet another contender as the key discipline by which all others are to be understood. Recently Plotkin[17] has argued forcefully that knowledge acquisition, human society, science, and all other academic disciplines can in a sense be regarded as biological phenomena. Darwinism therefore becomes the fundamental principle underlying all of human knowledge. Of course there are echoes in this of Popper's " evolutionary epistemology "[18], although for Popper science operated in a fashion parallel to evolutionary biology, and as an extension of it, rather than science being reducible to a biological phenomenon. Again, the theologians have resisted the claims of evolutionary thinking which " plays a certain kind of quasi-religious role in contemporary culture : it is a shared way of understanding ourselves at the deep level of religion, a deep

12. K.R. Popper, *The Poverty of Historicism*, op. cit.

13. See *Social Text*, 46-47 (1996).

14. I. Ker, " The Idea of a Catholic University ", in K. O'Dubhchair (ed.), *The Idea of a Catholic University in Mayo* : Newman Conference 1996, Mayo, 1997, 8-16.

15. A. Plantinga, " On Christian Scholarship ", in T.M. Hesburgh (ed.), *The Challenge and Promise of a Catholic University*, 1994.

16. I. Ker, " The Idea of a Catholic University ", op. cit.

17. H. Plotkin, *Darwin Machines and the Nature of Knowledge : Concerning Adaptations, Instinct and the Evolution of Intelligence*, London, 1994.

18. K.R. Popper, " Evolutionary Epistemology ", in D. Miller (ed.), *A Pocket Popper*, Fontana, 1983, 78-86.

interpretation of ourselves to ourselves, a way of telling us why we are here, where we come from, and where we are going "[19].

CONCLUSIONS

Scientists and the organisations which speak for them still generally hold a Mertonian view of science which has deep origins in the Nicene characteristics (one, holy, catholic and apostolic) of the Church from which Western science has sprung. This view entails a particular scientific morality, and the history of science has naturally been used as an ethical tool in support of that morality. However, this has led to a gap between how the history of science is used educationally by " ethicists " of science and how it is used by proponents of the " new " science studies.

I would like to suggest that the history of science can and should be used to support the ethics of science : but this should be done without falling into a form of the historicism that Popper condemned, and without failing to respond somehow to the influential strands of post-modern scholarship which would deny the very possibility of objective morality in science.

Ethics is an interdisciplinary field par excellence, and when we look at the ethics of science we should be mindful of the different disciplines which claim authority or have done so in the past. It is satisfying that the history of science itself provides an effective way of looking at those disciplines so that the fashionable claims of social theory and of Darwinistic philosophy today can be seen in proper historical perspective.

19. A. Plantinga, " On Christian Scholarship ", *op. cit.*

VERS UNE NOUVELLE MODÉLISATION DES RAPPORTS ENTRE SCIENCE ET POLITIQUE

Sébastien BRUNET

INTRODUCTION

Dans la mesure où les sciences constituent un certain pouvoir faire, une certaine maîtrise de la nature, notamment à travers leur rôle prédictif ; elles participent indirectement aux pouvoirs des êtres humains les uns sur les autres. On pourrait donc dire que les sciences sont des représentations de ce qu'il est possible de faire et par conséquent, des représentations de ce qui pourrait être l'objet d'une prise de décision dans nos sociétés post-industrielles[1]. De tout temps, un lien entre pouvoir politique et connaissance a été construit. Quelles relations peut-on envisager entre les sciences et les décisions politiques ou sociales ? On peut actuellement parler de politique technocratique lorsque l'on prend des décisions influencées, déterminées, ou légitimées par des recherches scientifiques. Mais, si dans les processus politiques traditionnels, la science apparaît comme un " allié " nécessaire à l'assise de la légitimité de la décision politique, il semble émerger une nouvelle tendance, poussant au devant de la scène, non plus une science universelle objective et sûre d'elle, mais une science hésitante et consciente de ses limites. En effet, le concept de science peut être déterminé de manières diverses, ce qui n'est pas sans influence sur la façon dont celle-ci intervient dans la sphère politique. Ce que l'on proposera donc dans ce texte, est une analyse des effets de la transformation du statut de la science, sur une redéfinition de certains modes de prise de décision. La bioéthique constitue à ce sujet un excellent terrain d'investigation puisque celle-ci se développe partout à travers nos sociétés post-industrielles. Plusieurs définitions de ce concept peuvent être défendues, mais on se limitera à une seule, déterminant la bioéthique comme une approche nouvelle pour la prise de déci-

---

1. Dans d'autres types de sociétés, la science ne constitue pas le mode de connaissance privilégié, et l'on choisira plutôt une connaissance mystique ou encore autoritaire. Sur ce sujet, voir B. Malinowski, *A scientific theory of culture and other essays*, Chapel Hill, 1944.

sion, des enjeux éthiques eux aussi nouveaux liés à l'utilisation croissante des technologies dans ce qui touche directement à la vie humaine et à la santé. Cette dernière définition est celle adoptée par Guy Bourgeault[2].

LE CONCEPT DE " POLITIQUE "

On définira celui-ci comme ce qui relève du domaine de la prise de décision, de l'action, visant à l'organisation de la vie collective dans un groupe d'hommes organisé. De plus, le concept de politique est intimement lié à la notion de " changement ". En effet, un des principaux défis auquel toute société humaine se voit confrontée est celui d'appréhender et de gérer le changement[3]. Il s'agit en réalité d'une recherche sécuritaire. C'est notamment ce que l'on observe lorsque l'homme aménage la nature selon ses besoins. Une autre caractéristique est le conflit à propos de ces aménagements, que faire, comment le faire et quand ? Ces conflits se déroulent non seulement sur les objectifs mais aussi sur les méthodes mises en oeuvre pour y parvenir. On se trouve ici au coeur de l'activité politique qui peut être trouvée à chaque fois que les individus sont engagés dans des processus de décision à propos de changements futurs. Le politique prend place aussi dans les processus de résolution de conflits à propos de méthodes à employer ou de buts à atteindre. Le pouvoir politique est donc ce qui permet de fixer la marche à suivre et d'imposer aux autres membres de la collectivité des règles de vie en commun[4].

LE CONCEPT DE " SCIENCE "

Introduction

Lorsque l'on parle habituellement de " science " dans nos sociétés, cela réfère presque aussitôt à des concepts tels que l'universalité, l'objectivité[5]. On a l'impression d'être en présence d'une activité sociale tout à fait différente des autres, une sorte de comportement humain purifié de toutes composantes culturelles ou sociales. Mais à côté de ce courant majoritaire, se développe une autre analyse du concept de " science ". On se limitera donc, à ces deux aspects généralement admis.

2. G. Bourgeault, " Qu'est-ce que la bioéthique? ", in *Les fondements de la bioéthique*, Textes réunis par Marie-Hélène Parizeau, Bruxelles, 1992, 27-47 (Sciences, Éthiques, Sociétés).

3. A. Renwick, I. Swinburn, *Basic Political Concepts*, London, Melbourne, Sydney, 1980.

4. Si l'on en croit Emile Durkheim, " L'État n'est rien autre chose que l'ensemble des institutions destinées à assurer le pouvoir d'une minorité sur une majorité. ", *Textes. 1. Éléments d'une théorie sociale*, Présentation de Victor Karady, Paris, 1975, Chapitre " Théories Allemandes de la société ", 344-391, 349.

5. A.F. Chalmers, *Qu'est-ce que la science ? Récents développements en philosophie des sciences : Popper, Kuhn, Lakatos, Feyerabend*, Paris, 1990 ; [orig. *What is this thing called science? An assessment of the nature and status of science and its methods,* St. Lucia, 1982].

Le concept de " science "

Deux principales visions des sciences, l'une positiviste et l'autre historique.

Dans la première, les sciences découlent pratiquement en ligne directe des observations. Elles sont en quête de *la vérité* scientifique. Les sciences découvrent les idées éternelles qui organisent le monde : les lois immuables de la nature. Les concepts scientifiques sont donc effectivement des concepts découverts dans la mesure où ils ne font que rejoindre ce que, de toute éternité, la nature incarnait. Dans cette perspective, les concepts ne sont pas des constructions visant à organiser notre vision, mais ils rejoignent une sorte de réalité en soi[6].

Dans l'approche historique, les représentations scientifiques découlent des représentations mythiques antérieures de la société. La théorie et le langage sont toujours là bien avant les observations effectuées par les hommes. Le travail scientifique sera un travail d'imagination, un travail d'invention, par lequel la communauté scientifique va remplacer certaines représentations par d'autres, trouvées plus adéquates selon nos projets humains[7]. Chaque discipline est donc une construction historique, conditionnée par une époque et des projets spécifiques. Les sciences apparaissent plus précisément comme les méthodes de description dont s'est dotée la " bourgeoisie "[8] lorsqu'elle a commencé à regarder le monde avec un autre regard, celui de l'étranger qui calcule. Voyons d'un peu plus près cette émergence des sciences modernes.

D'où vient la " science moderne " ?

L'on partagera l'analyse historique proposée par Gérard Fourez (1996), tout en reconnaissant avec son auteur qu'une telle démarche " est toujours une construction théorique (et donc idéologique) simplifiée "[9]. On peut considérer qu'il y a environ mille ans, et à peu près jusqu'au XIIe siècle, la vision qu'avaient du monde les occidentaux, était fortement liée à leur existence dans des villages autarciques. L'ensemble de la vie d'un individu se déroulait entièrement dans le même environnement humain. Le monde entier était humanisé,

6. Kant distingue " la chose en soi " (noumène) du " phénomène " ou chose phénoménale, c'est-à-dire ce que nous percevons et comprenons. Nous ne voyons le monde qu'à travers nous et nous n'avons jamais accès qu'à des phénomènes déjà structurés dans notre connaissance. La chose en soi échappe donc à la raison et à la perception.

7. Si les théories sont faites par les humains et pour les humains, il ne s'agit pas de tomber ici dans le relativisme, mais il devient possible de percevoir que dans notre histoire humaine, il y a place pour une variété de vérités, plutôt qu'une seule, si facilement totalitaire dans la mesure où l'on veut l'imposer à tous et en toute circonstance.

8. Il ne s'agit pas en cette occurrence du sens populaire du terme de " bourgeoisie " tel que l'a développé le marxisme, mais plutôt de celui caractérisant cette classe sociale apparue au Moyen Age qui est parvenue à remplacer l'aristocratie comme classe dominante.

9. G. Fourez, *La construction des sciences. Introduction à la Philosophie et à l'Éthique des Sciences*, Bruxelles, 1988 (Le point philosophique) ; *Idem,* 3e éd. rev., Paris, Bruxelles, 1996, 121 (Sciences, Éthiques, Sociétés)].

il n'existait pas d'objet purement matériel. C'est un monde que l'on ne domine pas mais que l'on essaie éventuellement d'apprivoiser, notamment grâce à la magie ou à la religion. Pour le marchand bourgeois, le monde " apparaît comme un lieu de plus en plus neutre et avec de moins en moins de structure humaine "[10]. Le centre du monde devient l'intériorité non plus du centre du village, mais celle liée à l'individu. On commence à faire la différence entre l'intérieur qui accompagne toujours l'individu et est subjectif et l'extérieur considéré comme un objet. L'Occident s'est donc créé des méthodes de description (des technologies intellectuelles) telles que ce qui a été observé à un endroit du globe peut être rapporté à autre. L'objectivité paraît ainsi comme une manière détachée de voir le monde de ce qui le compose dans sa globalité. L'objectivité n'existerait pas en elle-même, mais serait la production d'une culture. Pour comprendre une description scientifique, il faut avoir une culture scientifique, c'est-à-dire les prérequis nécessaires à l'interprétation et à la compréhension des concepts. Un univers conceptuel mental, intériorisé par chaque scientifique, va remplacer l'univers partagé des villages. Les scientifiques sont des gens qui se comprennent pour avoir uniformisé leur perception du monde, exactement comme le font les habitants d'un même village. Avec la mentalité bourgeoise, on essaie de maîtriser et donc de prédire son environnement. Les sciences modernes sont ainsi liées à l'idéologie bourgeoise et à sa volonté de maîtriser le monde et de contrôler l'environnement.

Une critique de la vision positiviste des sciences : La sociologie des sciences

Les sciences sont un phénomène de société. Tout d'abord, les sociologues se sont intéressés à ce qu'il y a autour des sciences. On pense par exemple à l'analyse des liens existants entre les scientifiques et d'autres institutions sociales, aux influences diverses du financement de la recherche scientifique par les intérêts industriels ou militaires. En 1962, Th. Kuhn[11] (et sa notion de paradigme) commence à s'intéresser à l'influence des éléments sociaux sur la structure des connaissances scientifiques. Toute discipline scientifique est selon lui déterminée par un paradigme. Ces paradigmes sont culturellement et historiquement construits. L'objet d'une discipline ne préexiste pas au paradigme, il est déterminé par celui-ci. La science normale essaie de résoudre les problèmes à l'intérieur du paradigme et tire de lui ses questions et la traduction de ses réponses. En cas d'inadéquation entre un paradigme et les demandes, on peut entrer en période de révolution scientifique. L'aspect institutionnel des paradigmes était ainsi mis en évidence. On commence donc à percevoir que les contenus mêmes des sciences sont structurés autour de projets, de préjugés, et

10. G. Fourez, *La construction des sciences, op. cit.*, 123.
11. T.S. Kuhn, *La structure des révolutions scientifiques*, trad. Laure Meyer, Paris, 1983. [orig. T.S. Kuhn, *The Structure of scientific revolutions*, Chicago, London, 1962].

même de dominations sociales que l'on peut étudier. Mais on considérait toujours qu'au centre du travail scientifique, il y avait des éléments qui représentaient une objectivité absolue. Finalement, seule la rationalité scientifique restait à l'abri des recherches en sociologie : elle ne dépendait que de la raison pure. En 1973, Merton[12] s'est intéressé plus directement aux pratiques scientifiques. Il s'est focalisé sur la sociologie même de la communauté scientifique. Mais on ne considère encore en rien les contenus scientifiques. Enfin, ce dernier bastion fut attaqué. Des personnes comme Feyerabend[13] et Bloor[14] estiment que les contenus mêmes des sciences apparaissent comme des créations humaines par et pour des humains. Selon les sociologues des sciences, l'objectivité éternelle des observations scientifiques, n'apparaît éternelle qu'à cause de l'accoutumance à un certain nombre de présupposés et de catégories utilisées. Les sciences ont une objectivité relative, c'est-à-dire qu'elles ont une manière extrêmement efficace de mettre de l'ordre dans notre perception, dans notre monde, et de communiquer le type d'ordre qu'ensemble nous pouvons et voulons utiliser. " La science s'affirme aujourd'hui science humaine, science faite par des hommes et pour des hommes. La science suppose un enracinement social et historique et une interprétation globale qui n'est pas sans influence sur les recherches locales. Les scientifiques appartiennent à la culture à laquelle ils contribuent à leur tour "[15].

L'observation n'est donc pas purement passive, il s'agit plutôt d'une certaine organisation de la vision. En observant quelque chose, je le décris en utilisant toute une série de notions. Celles-ci se réfèrent toujours à une représentation théorique, généralement implicite. Observer c'est donc interpréter, c'est-à-dire intégrer une certaine vision dans la représentation théorique que l'on se fait de la réalité. Les scientifiques sont les participants d'un monde culturel et linguistique dans lequel ils insèrent leurs projets individuels et collectifs. Dire que quelque chose est objectif, c'est le situer dans un univers commun de perception et de communication, dans un univers conventionnel, institué par une culture. L'objectivité comprise ainsi n'est pas absolue mais toujours relative.

On peut donc considérer les sciences comme des technologies intellectuelles destinées à fournir des interprétations d'un monde qui correspondent à nos projets.

Les périodes paradigmatiques ont un enjeu de pouvoir social important : lorsque les disciplines se sont imposées, ont durci leurs concepts, ont gommé leurs origines sociales, les chercheurs jouissent d'une relative indépendance face au contexte social dans lequel ils évoluent. Cette indépendance est selon

12. R.K. Merton, *The Sociology of Science : Theoretical and Empirical Investigations*, edited and with an Introduction by N.W. Storer, Chicago, London, 1973.

13. P. Feyerabend, *Contre la méthode : Esquisse d'une théorie anarchiste de la connaissance*, trad. B. Jurdant et A. Schlumberger, Paris, 1979.

14. D. Bloor, *Sociologie de la logique ou les limites de l'épistémologie*, Paris, 1983.

15. I. Prigogine, I. Stengers, *La nouvelle alliance : métamorphose de la science*, Paris, 1980.

certains auteurs une condition essentielle au travail scientifique. Car, s'ils ne jouissaient pas de celle-ci, les scientifiques seraient totalement bloqués dans leur oeuvre de création. D'ailleurs, la proposition de Kuhn préserve l'essentiel pour les scientifiques : l'autonomie d'une communauté scientifique par rapport à son environnement socio-politique. Cette proposition institue l'autonomie comme norme et condition de possibilité pour l'exercice fécond d'une science. Le scientifique est devenu vecteur d'une créativité que n'aurait peut-être pas inspirée une attitude lucide, c'est-à-dire sceptique, quant au pouvoir des théories. Ainsi, l'autonomie d'une communauté travaillant sous paradigme est une condition indispensable au progrès scientifique.

Dans la mesure où les sciences sont toujours un certain pouvoir faire, une certaine maîtrise de la nature, elles sont liées par ricochet aux pouvoirs des êtres humains les uns sur les autres. Les sciences sont toujours des représentations de ce qu'il est possible de faire et par conséquent, des représentations de ce qui pourrait être l'objet d'une décision dans la société. De tout temps, un lien entre pouvoir politique et connaissance a été supposé. Quelle est la nature des relations que l'on peut envisager entre les sciences et les décisions sociales ? On peut parler de politique technocratique lorsque l'on veut prendre des décisions politiques influencées, ou déterminées, ou légitimées par des recherches scientifiques. Il s'agit d'une politique par la science.

RELATION ENTRE SCIENCE ET POLITIQUE

Cette relation a été modélisée par Jurgen Habermas[16] qui classe en trois catégories les manières de percevoir les interactions entre science et politique : les interactions technocratiques, les interactions décisionnistes, les interactions pragmatico-politiques.

Ces catégories n'existent pas à l'état pur, ce ne sont que de simples modèles conceptuels afin d'aider à l'appréhension et à la compréhension de la réalité.

Pour le modèle technocratique, les décisions reviennent aux experts. Dans nos sociétés post-industrielles, le modèle technocratique est fort répandu, car on croit facilement que notre observation scientifique du monde a une objectivité absolue. On gomme ainsi la particularité de notre vision, de notre société, et on aboutit à une société technocratique où l'on voudra fonder ou légitimer des décisions socio-politiques ou éthiques sur des raisonnements scientifiques prétendus neutres et absolus. On présuppose ainsi que le commun des mortels ne peut rien comprendre et l'on tend alors à s'en remettre à ceux qui savent, c'est-à-dire aux experts[17]. Ce seraient alors les connaissances qui détermine-

---

16. J. Habermas, *La technique et la science comme " idéologie "*, Paris, 1973.

17. Sur le concept d'expert, voir notamment, P. Lascoumes, *L'éco-pouvoir : environnements et politiques*, Paris, 1994 ; B. Latour, *Nous n'avons jamais été modernes : Essai d'anthropologie symétrique*, Paris, 1991 ; Y. Merchiers, P. Pharo, " Éléments pour un modèle sociologique de la compétence d'expert ", *Revue Sociologie du Travail*, n° 1 (1992) ; D. Nelkin, *Controversy, Politics of Technical Decisions*, 2nd ed., London, 1986.

raient les politiques à suivre, que celles-ci découlent directement de l'argumentation scientifique, ou qu'elles soient prises en son nom. On appelle donc technocraties, les systèmes politiques qui légitiment leurs décisions socio-politiques sur base d'éléments scientifiques. L'exemple fourni par les économistes en est une admirable illustration. Ceux-ci se posent en spécialistes de la conduite économique des pays en brandissant une légitimité scientifique qui d'ailleurs ne fait pas toujours l'unanimité au sein de la communauté scientifique. En matière économique, que cela soit au niveau des entreprises ou des États, les économistes constituent donc une caste de spécialistes qui étayent les grandes orientations ou décisions politiques[18]. On s'aperçoit donc ici de l'importance que recouvre la définition même du concept de science. Si l'on adopte une définition historique de celle-ci, l'approche technocratique commet un abus de savoir, parce que l'on admet que les connaissances scientifiques elles-mêmes ne sont pas neutres. Elles ont été construites selon un projet organisateur et ce dernier peut déterminer leur nature. La manière dont les experts se mettront d'accord relèvera plus de la logique d'une négociation socio-politique que d'un cadre de rationalité bien défini. Le statut de l'expert présente une ambiguïté fondamentale, car on demande à l'expert de décider en fonction de son savoir scientifique. Or ce savoir dépend, comme tel, d'un paradigme et n'est applicable, dans le sens strict, que d'après les conditions définies par ce paradigme et par le laboratoire qui y est lié. Mais l'avis d'expertise demandé est, quant à lui, destiné à la vie quotidienne. Ce qui revient à poser une question d'ordre social ou économique à l'expert. L'expertise est donc liée à la manière dont l'expert traduit le problème de la vie courante dans son paradigme disciplinaire. Bref, il est inexact de croire que c'est uniquement au nom de sa discipline que l'expert " sort de son laboratoire "[19].

Selon le modèle décisionniste, l'expert demandera au " client " quels sont ses objectifs. Une fois connues les finalités et les valeurs du client, l'expert, grâce à ses connaissances, trouvera les moyens les plus adéquats pour atteindre ces objectifs. Ce modèle distingue donc entre décideurs et techniciens. Les uns déterminent les fins, les autres les moyens. Max Weber[20] a lié cette manière de voir avec une théorie de la rationalité. Un plan d'action est rationnel quand les moyens correspondent aux objectifs choisis. Ces objectifs ne peuvent être choisis rationnellement, puisqu'ils relèvent des valeurs. Le lieu de la rationalité serait alors le choix des moyens. Pour Weber la science doit être totalement distincte du politique, de peur que ce dernier ne vienne la corrompre. Il se battait contre ceux qui risquaient de souiller la pureté de la pensée rationnelle en y mêlant des prises de positions politiques. En effet, pour Weber la vérité est

18. Les pays membres de l'Union Européenne sont directement confrontés à ce type de décision, dans la mise en oeuvre de la monnaie unique et de ses critères d'admission.

19. I. Stengers, *Sciences et pouvoirs : la démocratie face à la technoscience*, Paris, 1997.

20. M. Weber, *Essais sur la théorie de la science*, Paris, 1965.

l'objet de la science, et les valeurs inhérentes au politique ne doivent pas intervenir dans cette démarche. Ce qui semble préoccuper Weber[21], c'est plus
l'intervention de la politique dans l'oeuvre scientifique que l'inverse, puisqu'il
reconnaît à la science un rôle primordial d'information auprès des décideurs
politiques. La science met à notre disposition un certain nombre de connaissances qui nous permettent de dominer techniquement la vie par la prévision,
aussi bien dans le domaine des choses extérieures que dans celui de l'activité
des hommes. La science contribue à une oeuvre de clarté. En présence de tel
problème de valeur qui est en jeu, on peut adopter pratiquement telle position
ou telle autre. Quand on adopte telle ou telle position il faudra, suivant la procédure scientifique, appliquer tels ou tels moyens pour pouvoir mener à bonne
fin son projet. Le professeur peut seulement vous montrer la nécessité de ce
choix, mais il ne peut faire davantage. Il peut également vous indiquer que
lorsque vous voulez telle ou telle fin, il faudra consentir à telles ou telles conséquences subsidiaires qui en résulteront. Les savants peuvent encore vous dire
que tel ou tel parti que vous adoptez dérive logiquement, et en toute convic-
tion, quant à sa signification, de telle ou telle vision dernière et fondamentale
du monde. La science vous indiquera qu'en adoptant telle position vous servi-
rez tel dieu et vous offenserez tel autre parce que, si vous restez fidèles à vous-
mêmes, vous en viendrez nécessairement à telles conséquences internes, dernières et significatives. La science n'est pas là pour effectuer des choix mais
pour les éclairer. L'intérêt de ce modèle est de laisser du pouvoir aux non-spé-
cialistes tout en reconnaissant qu'il y a effectivement, à partir du moment où
les sciences sont devenues assez complexes, deux classes de citoyens : ceux
qui savent et ceux qui ne savent pas. Cette différenciation des rôles dans nos
sociétés est très poussée. Ce modèle permet aux gens de choisir en fonction de
leurs valeurs, ce qui en fait un modèle plus démocratique. Il laisse donc aux
gens leur mot à dire par rapport aux valeurs qu'ils poursuivent. Par contre, sa
faiblesse est de présupposer qu'une fois que l'on a déterminé les finalités, le
choix des moyens reste indifférent. Ce qui revient à ignorer que le choix des
moyens techniques détermine toute une organisation sociale et n'est donc pas
indifférent par rapport aux valeurs et aux fins poursuivies. Il semble donc que
le modèle décisionniste soit un leurre, et que les vertus démocratiques de celui-
ci aient finalement peu de poids face à l'importante inertie qu'entraîne la mise
en oeuvre de moyens réputés " techniques ". Plus on se trouve en présence de
technologies complexes, plus les lignes d'action raisonnables sont déterminées
par les technologies elles-mêmes et par conséquent devront être définies par
des spécialistes. Car une technologie n'est pas seulement un ensemble d'éléments matériels, mais aussi un système social. Les choix technologiques vont

---

21. Max Weber a tenté d'obtenir du pouvoir politique de son époque une reconnaissance politique. Mais en vain, car celle-ci ne lui a jamais été accordée. Ceci pourrait peut-être expliquer cette
position très tranchée qu'il soutient au sujet de la distinction entre le métier de scientifique et celui
du politique. Voir M. Weber, *Le savant et le politique*, 1959.

donc déterminer le type de vie sociale d'un groupe. Face à la possibilité d'un conditionnement de l'existence individuelle et sociale par les technologies, deux idées s'opposent : 1) les technologies sont à la disposition des hommes, ceux-ci ayant à décider, selon leur éthique, de la manière dont ils vont les utiliser. 2) Les technologies détermineraient même la vie. Elles télécommandent les structures des sociétés qui les adoptent. Elles véhiculent déjà des structures de société.

Dans le concret, les gens s'aperçoivent souvent qu'ils ont à s'adapter, bon gré mal gré, aux technologies et que celles-ci dictent finalement la manière dont ils doivent vivre et travailler.

Quant au modèle pragmatico-politique, celui-ci suppose une perpétuelle négociation entre le technicien et le non-technicien. Ici, l'interaction entre les experts et les non-experts se fait sans arrêt. Il n'y a pas de limites prédéfinies au niveau des objectifs et des moyens, car tout est systématiquement négociation. Ce modèle ne considère pas la technique comme un pur instrument mais aussi comme une organisation et une attitude face à des questions humaines. Il existe donc une multiplicité de langages, de grilles d'analyse différentes pouvant s'appliquer à une situation particulière. De cette manière, on reconnaît que l'ensemble des personnes concernées ont un avis à donner, tant au niveau de la définition de la problématique que dans la résolution de celle-ci. Ce modèle permet également d'être sensible à un concept nouveau, celui d'" être hybride " développé par Bruno Latour[22]. Certains phénomènes sociaux ne sont plus strictement compartimentés, mais " voyagent " entre différentes sphères (politique, scientifique, sociale). C'est la nature même de ces phénomènes qui semble être à l'origine de cette hybridation.

Il apparaît ainsi que les débats entre techniciens et non-techniciens sont fondamentaux dans ce modèle. C'est dans cette perspective que la vulgarisation scientifique prend une grande importance. Deux manières de comprendre la vulgarisation scientifique : Premièrement, comme une opération de relations publiques de la communauté scientifique vis-à-vis du bon peuple. Le but n'est évidemment pas de transmettre une véritable connaissance, mais de donner un certain vernis de savoir (connaissance factice). Dans la seconde perspective, la vulgarisation scientifique vise à donner aux gens un certain pouvoir. Dans une société fortement basée sur les sciences et les technologies, la vulgarisation scientifique a donc un enjeu socio-politique assez important. Dans la seconde perspective, la vulgarisation est une transmission de pouvoir. Pour être un individu autonome et un citoyen participatif dans une société hautement technicisée, il faut être scientifiquement et technologiquement alphabétisé. Et dans les deux cas, cet exercice de communication de la communauté scientifique à des-

---

22. B. Latour, *Nous n'avons jamais été modernes..., op. cit.* ; B. Latour, " Esquisse d'un parlement des choses ", *Écologie politique : sciences, culture, société*, n° 10 (1994).

tination des profanes, aura pour conséquence un renforcement du pouvoir symbolique détenu par celle-ci.

A quel(s) modèle(s) nos sociétés correspondent-elles le mieux ? Il semble difficile dc répondre à cette question, dans la mesure où ces modèles n'existent pas en tant que tels dans la réalité, mais sont de simples instruments de compréhension des interactions entre la science et le politique. Toutefois, il semble que le modèle technocratique soit le plus répandu dans nos sociétés post-industrielles. En outre, un élément essentiel transparaît à travers ces trois cas de figure, à savoir, l'existence d'une relation privilégiée entre la science et le politique qui a pour finalité la légitimité des décisions politiques. En fait, on voit ici que la question centrale est celle de la légitimité de la décision politique.

Dans une technocratie, la décision incombe en réalité à ceux qui savent, c'est-à-dire aux experts. Ils sont donc les véritables décideurs politiques, même si formellement la décision émane d'autres instances. Dès lors, les détenteurs institutionnels du pouvoir politique n'assument plus qu'une fonction d'entérinement des décisions scientifiques, à moins que représentants politiques et scientifiques ne soient les mêmes personnes. Pour que l'action politique puisse exister, il faut que celle-ci soit acceptée par ses destinataires. Dans la perspective d'un système technocratique, la légitimité sur laquelle s'appuient les décideurs politiques, n'émane pas des élections, mais bien de la confiance que porte la communauté en la science. Dès lors, la définition de cette dernière devient un enjeu politique dans la mesure où il est plus facile d'imposer des décisions prises sur base d'une science universelle et objective plutôt que sur base d'une activité humaine comme les autres.

Dans le modèle décisionniste, la prise de décision politique a deux composantes, qui bien que complémentaires ne se fondent pas sur la même légitimité. La première composante d'une telle décision se situe au niveau de la détermination des objectifs (valeurs) par ce que l'on appelle communément " le politique ". Cette fraction de la décision politique " globale " est en principe inaccessible aux scientifiques. Sa légitimité est généralement fondée sur les modes de représentation politique. La deuxième composante de la décision politique, se situe au niveau des moyens à mettre en oeuvre pour atteindre ces objectifs. C'est ici que les scientifiques interviennent dans les limites préalablement définies par le politique. Leur légitimité dépend de la représentation de la science que l'on a dans la société. Il existe donc deux types de légitimité soutenant une même décision, l'une politique, l'autre scientifique.

Dans le modèle pragmatico-politique, les décisions sont des objets qui voyagent entre les sphères politique et scientifique. Il s'agit d'un processus complexe d'interactions dans lequel s'opère une redéfinition constante des objectifs et des moyens. Dans ce cas de figure, la légitimité des décisions politiques repose justement sur cet échange et cette renégociation constante entre science et politique. Cette dynamique suppose une représentation historique de la science, puisque celle-ci remplit une fonction de partenaire à part entière. Les scientifiques discutent et rediscutent des diverses solutions envisageables sur le même pied d'égalité que leurs interlocuteurs politiques.

LA TRANSFORMATION DU STATUT DE LA SCIENCE ET L'ÉMERGENCE D'UN NOUVEAU TYPE D'OUTIL D'AIDE À LA DÉCISION POLITIQUE

Introduction

Même si la définition traditionnelle de la science représentée par le courant positiviste, rallie encore aujourd'hui une majorité d'opinions, il semble que peu à peu celui-ci perde de sa crédibilité pour faire place à une conception de la science plus " relativiste ", telle que l'approche historique nous la présente. Deux principaux éléments semblent étayer ce changement. D'une part, le développement croissant du courant de la sociologie des sciences, qui remet en question la représentation classique de la science et l'étudie en tant qu'activité humaine comme les autres, ni plus objective, ni plus universelle que les autres. D'autre part, cette tendance semble renforcée par l'actualité scientifique, à travers laquelle on aperçoit une science plus incertaine et moins catégorique à propos des conséquences et de la maîtrise de ses produits. Là où certains voient une réduction de l'incertitude technique par la science, d'autres voient une augmentation de l'incertitude sociale. C'est en réalité cette dernière qui constitue l'élément essentiel du point de vue du spécialiste en sciences administratives. En effet, si les technologies modernes parviennent à donner quelques réponses techniques à certaines questions soulevées par la société, il en est d'autres auxquelles les sciences ne sont pas à même de satisfaire et qui ajoutent aux interrogations techniques des incertitudes sociales. Cette situation entraîne une difficulté pour les décideurs politiques à fonder leurs décisions ou certaines parties de celles-ci sur des arguments scientifiques. Ceci explique l'émergence dans certains domaines, vraisemblablement plus sensibles que d'autres, de nouveaux modes de prise de décision politique. On pense plus particulièrement au domaine de la bioéthique comme idéal-type de ce phénomène.

La bioéthique

Le concept de " bioéthique " s'est progressivement développé pour devenir aujourd'hui incontournable dans la gestion de nos sociétés démocratiques post-industrielles. Cette nouvelle notion est intimement liée au développement technologique et aux conséquences de celui-ci sur la société civile. Deux aspects intéressants de la bioéthique peuvent être soulevés. D'une part, elle opère une relation entre le monde scientifico-technique et le monde politique (caractérisé par l'action). D'autre part, elle met en exergue la difficulté pour nos systèmes démocratiques de gérer certains types de conflits et propose un nouveau mode de discussion et de processus décisionnel.

Au milieu des années soixante, les Américains prennent conscience des problèmes soulevés par le développement des nouvelles technologies biomédicales. En effet, celles-ci remettent en question d'importants principes portants tant sur la nature de la vie et de la mort que sur la responsabilité politico-éthique des scientifiques inhérente au progrès technologique. Or, l'éthique médicale de l'époque, essentiellement catholique, éprouve quelques difficultés d'adaptation face à ces bouleversements politico-technologiques. Selon Leroy Walters[23] du Kennedy Institute of Ethics, l'élément déclencheur dans la transformation de l'éthique consiste dans la prise de position de Paul VI répétant, en 1968 et malgré les avis contraires d'une commission *ad hoc*, l'opposition traditionnelle de l'Église catholique aux méthodes contraceptives. Cet événement marque définitivement l'incapacité de la théologie catholique à intégrer le présent. Dès lors, la nécessité s'imposait de créer des structures, hors de la dominance ecclésiastique, afin de faciliter la recherche et le dialogue entre biologie, médecine et éthique[24].

On peut définir la bioéthique de trois manières différentes[25]. D'une part, en tant que simple mise à jour de l'ancienne morale médicale ; d'autre part, en la considérant comme une éthique de la vie qui embrasse tout (une nouvelle discipline) ; et enfin, comme une approche nouvelle pour la prise de décision, des enjeux éthiques eux aussi nouveaux liés à l'utilisation croissante des technologies dans ce qui touche directement à la vie humaine ou à la santé, ou dans le champ des pratiques biomédicales. C'est cette dernière perspective, définissant la bioéthique en termes d'outil d'aide à la décision et d'instrument de conciliation du progrès technologique avec la morale, que l'on retiendra. La forme institutionnelle que prend ce type de processus décisionnel est le comité de bioéthique.

CARACTÉRISTIQUES DE CE NOUVEL OUTIL D'AIDE À LA DÉCISION POLITIQUE

La bioéthique : un être hybride

L'hybridation, telle qu'on la définira ci-dessous, constitue une des traces identifiables du changement de statut de la science. Il s'agit d'un concept sociologique nouveau définissant des phénomènes dont l'ampleur ne se cantonne plus à une sphère particulière, mais bien au contraire, concerne maints domaines de nature diverse. Autrement dit, le mot-clef est ici la non-compartimentation des phénomènes sociaux. En effet, des problématiques, telles que celles que la bioéthique soulève, ont des conséquences aussi importantes que

23. L. Walters, " Religion and the Renaissance of Medical Ethics in the United States : 1965-1975 ", in E.E. Shelp (ed.), *Theology and Bioethics : Exploring the Foundations and Frontiers*, Dordrecht, Boston, Lancaster, Tokyo, 1985, 3-16 (Philosophy and Medicine, vol. 20).

24. An interview with Daniel Callahan, " Beyond Individualism : Bioethics and the Common Good ", *Second Opinion*, 9 (1988), 52-69.

25. Voir G. Bourgeault, " Qu'est-ce que la bioéthique ? ", *op. cit.*, 27-47.

diversifiées dans les sphères scientifique, morale, économique et politique. L'identification d'un phénomène à une seule sphère de référence n'est plus possible actuellement dans nombre de situations. La distinction classique opérée dans nos sociétés démocratiques post-industrielles entre société et science, fut ces dernières années remise en question par une multitude de situations échappant à cette classification rigoureuse. Autrement dit, les débats passent aujourd'hui à l'intérieur même des sphères scientifique et politique et non plus entre celles-ci. Ainsi, certaines problématiques ne sont plus traitées comme " purement " scientifiques, mais aussi pleinement politiques. Le raisonnement inverse, passant du politique au scientifique est également envisageable. Précisément, dans ces deux cas de figures, le développement technologique constitue un idéal-type. En effet, si une découverte scientifique concernant le génome humain permet de traiter préventivement certaines maladies héréditaires, ce progrès technologique deviendra à proprement parler un débat politique. Pour poursuivre l'explication dans la relation inverse, une décision politique ayant trait à la bioéthique ne peut s'envisager sans susciter aucun débat scientifique. Ainsi, on ne fait pas que se renvoyer la balle, on la garde, on la regonfle et on la travaille.

Cette hybridation a pour conséquence directe une autre caractéristique essentielle que l'on retrouve dans la mise en oeuvre du concept de bioéthique, à savoir, l'interdisciplinarité[26]. Lors de la mise en place de comités de bioéthique partout à travers le monde, on retrouve cette même constante qu'est la mise en commun de savoirs spécialisés mais différents pour débattre de problèmes particuliers. Quelle interprétation donner de ce phénomène et surtout, pourquoi une telle traduction de l'hybridation au niveau institutionnel ?

L'interdisciplinarité

L'approche du monde par une discipline particulière est biaisée et généralement trop courte. On prend ainsi conscience qu'une question déterminée peut requérir une multiplicité d'approches. D'où le concept d'interdisciplinarité.

L'on rencontre habituellement deux philosophies à l'égard de l'interdisciplinarité : 1) en rassemblant plusieurs approches, on espère une superscience, super objective. On pense ainsi échapper au cadre strict déterminé par l'utilisation d'un seul paradigme. 2) pratique concrète de négociations pluridisciplinaires, face aux problèmes concrets du quotidien.

Dans la pratique cela se traduit par deux attitudes très différentes. La première perspective espère qu'une approche interdisciplinaire construira une nouvelle représentation du problème qui sera beaucoup plus adéquate dans l'absolu. Mais en créant une sorte de superscience plus objective que les autres, on ne fait que produire une nouvelle approche, une nouvelle discipline.

26. Sur le thème de l'interdisciplinarité, voir Ph. Roqueplo, *Entre savoir et décision, l'expertise scientifique : une conférence-débat organisée par le groupe Sciences en question*, Paris, 1997.

En réalité, il ne s'agit dans ce cas que d'une construction d'un nouveau paradigme. On se fourvoie donc en croyant obtenir une superscience sur base d'une
addition de paradigmes différents. On est alors en présence d'une technocratie
interdisciplinaire.

La seconde perspective ne considère pas l'interdisciplinarité comme une
sorte de nouveau discours qui serait au-delà des disciplines particulières, mais
elle y est vue comme une pratique spécifique en vue de l'approche des problèmes de la vie quotidienne. Dans cette dernière acception, l'interdisciplinarité
est perçue comme une pratique essentiellement politique, c'est-à-dire comme
une négociation entre différents points de vue pour finalement décider d'une
représentation considérée comme adéquate en vue d'une action. On est dès lors
face à un modèle de type pragmatico-politique.

Dans la pratique, il semble que le fonctionnement des comités de bioéthique
corresponde plus à cette deuxième conception de l'interdisciplinarité, qu'à la
mise en oeuvre d'une superscience. La principale justification de cette approche particulière de la bioéthique, tire son fondement de la nature même des
problèmes traités par celle-ci. Ce sont des problématiques hybrides qui néces-
sitent une approche différente de celle proposée par les processus technocratiques de prise de décision.

CONCLUSION

Quelle analyse peut-on faire au sujet de ces modèles d'interaction, entre la
science et le politique, dans nos sociétés démocratiques occidentales ? Le
modèle décisionniste semble peu intéressant, puisqu'il comporte une faiblesse
relativement importante, en sous-estimant l'influence que peut avoir le choix
des moyens sur la détermination des buts à atteindre. La technocratie est quant
à elle très répandue et peut se retrouver en filigrane dans quantité de décisions
politiques. Ce modèle laisse peu de place à la participation citoyenne en per-
mettant aux experts ou aux politiques de décider sous le couvert de la rationalité scientifique. Toutefois, la technocratie dans certaines circonstances s'efface
au profit du modèle pragmatico-politique. Le développement de celui-ci doit
être mis en parallèle avec l'évolution récente du statut de la science et donc de
la représentation de celle-ci dans la société. Le phénomène d'hybridation, dont
on a parlé plus haut, implique une gestion différente d'un certain type de questions ou de problématiques nouvelles. Ce mode de décision se caractérise par
une interrelation constante entre les sphères politique et scientifique, que l'on
observe dans les comités interdisciplinaires de bioéthique. Dans certains de
ceux-ci, comme au Danemark, les citoyens participent directement aux travaux. D'autres par contre, comme au Canada[27], recueillent lors de la prépara-
tion de leurs documents les commentaires des citoyens intéressés avant de

27. G. Durand, *La bioéthique : nature, principes, enjeux*, Paris, 1989.

remettre leur rapport final au gouvernement. Enfin, dans la majorité des autres comités de bioéthique, on retrouve, soit des politiques et des scientifiques, soit uniquement des scientifiques. Quant aux citoyens, ils ne sont pris en considération ni directement, ni indirectement. Toutefois, certains soutiennent que lorsque l'on intègre des mandataires politiques, les citoyens sont représentés. Or, les mandataires politiques sont, avant d'être les représentants des citoyens, des experts de la prise de décision, et ils ne peuvent dès lors se substituer aux experts de la vie de tous les jours que sont les citoyens. Enfin, selon la nature des matières concernées par la prise de décision, on adoptera plus ou moins facilement une procédure technocratique ou pragmatico-politique. On a montré qu'en matière de bioéthique la préférence allait au modèle pragmatico-politique. Ainsi, les citoyens, destinataires des décisions politiques, seront rassurés face à une adaptation du mode de décision, en fonction des caractéristiques du domaine sur lequel porte la décision. Néanmoins, si ce type de processus décisionnel semble séduisant, il n'apparaît pas toujours intéressant en toute matière. En effet, dans certaines circonstances, les décisions doivent se prendre dans des laps de temps très restreints, ce qui fait du modèle technocratique un instrument beaucoup plus efficace. Toutefois, l'évolution récente des technologies modernes semble indiquer que de plus en plus de procédures de type pragmatico-politique verront le jour, afin de répondre efficacement aux incertitudes sociales grandissantes de nos sociétés.

CONCEPT OF PROGRESS AND GEOGRAPHICAL THOUGHT

WITOLD J. WILCZYŃSKI

> " Something is wrong with science —
> fundamentally wrong. (...) It has been my
> sad observation that by mid-career there
> are very few professionals left truly work-
> ing for the advancement of science, as
> opposed to the advancement of self ".
>
> T. van Flandern[1]

In approaching the question of progress and geography, I might have been expected, as a professional geographer, to think of some famous geographers' contributions to technological and cultural advancement. However, keeping in mind the present-day situation in Polish geography, it might have been thought more appropriate for me to concentrate on the very meaning of the word progress.

One of the permanent features of the history of Western civilization is the aspiration for progress. This is the notion which has continued to appear in various contexts in all areas of culture. First of all, it has been utilized as a synonym for scientific and technological advancement. And because both science and technology have been thought to improve greatly the standards of life, the notion of progress, in its scientific-technological sense, has gained an almost univocal positive connotation in social consciousness. Until now, progress is considered to be something positive and good, as opposed to such terms like stagnation, backwardness, or recession. Such a positive value of progress is not limited to the scientific-technological sphere. In Polish tradition there appeared the idea of social progress, identified with the spread of the leftist ideology, and particularly, with the Marxist-Leninist doctrine. Ideologists had utilized

---

1. T. van Flandern is an American astronomer and methodologist. The sentence is taken from his book entitled *Dark Matter, Missing Planets and New Comets : Paradoxes Resolved, Origins Illuminated*, Berkeley, California, 1993, xv-xvii.

WITOLD J. WILCZYŃSKI

the " progressivist " phraseology to discredit all conservative ideas calling them " bourgeois ", " reactionary ", or simply " American ".

There are repeated statements in which the notion of progress have been ideologically abused. As a result, this notion has greatly influenced not only the ordinary people's way of thinking, but also the way of argumentation in scientific research. The general aspiration for being " progressive " has been shared by the most of research workers in all fields of knowledge. The over-throw of communism in the last few years has however caused progress, like many other general notions, to become the object of reconsideration and con-testation.

At present, the idea of progress is in the state of crisis both in Eastern Europe and in the West[2]. Its source is the general distrust of technological innovation based on quasi-religious reverence for pre-technological nature. This is in fact part of an age-old Western cultural heritage. This distrust was even represented by Ancient Greeks (Parmenides, Heraclitus), and Romans (Horace), as well as by numerous modern thinkers. J.-J. Rousseau has written that " progress of the sciences and arts has added nothing to our real happiness (...) and has corrupted our morals "[3]. This assertion has been repeated in writ-ings by romantic transcendentalists, creators of environmental movements and authors of various eco-philosophies. According to Andrew N. Woznicki, Phi-losophy Department, University of San Francisco, progress is made at the expense of the traditional value system and leads young Californians to a com-plete isolation of the individual from his social establishment, to a radical sep-aration from any higher values and reducing the spiritual to bodily needs, and to a tragic feeling of an ecological threat of losing any contact with the natural world[4]. Nowadays even economic progress has been called an illusion since it does not take into account the depreciation of natural capital, including non-renewable resources[5].

According to Martin Heidegger, the world is dominated by the technologi-cal *Gestell*. This technological framework in which we all live today, he regards as a danger to human existence (" only a God can save us "). Contem-porary thinkers have shown that there exists no synchrony between technolog-

2. Extensive criticism in relation to progressivism is expressed in the article by W.L. McBride, " The Progress of Technology and the Philosophical Myth of Progress ", *Philosophy and the His-tory of Science : A Taiwanese Journal*, 1, n° 1 (1992), 31-58. From the point of view of environ-mentalism the same questions were discussed during the conference held in Lublin in 1992. Results of the conference were published in the special volume *Crisis of the progress concept — an ecological dimension (Kryzys idei postępu - wymiar ekologiczny)*, in Stanisław Kyć (ed.), 1993.

3. J.-J. Rousseau, *Discours sur les sciences et les arts. Du Contrat Social*, Paris, 1954, 22.

4. A.N. Woznicki, " New Spirituality in California ", *Dialogue and Humanism* (Summer 1991), 18-19. The peculiar weakness of our technological civilization is shown in writings by the Polish philosopher, creator of the eco-philosophy, Henryk Skolimowski.

5. L.R. Brown, " The Illusion of Progress ", in L.R. Brown, W.W. Norton & Company (eds), *State of the World 1990*, New York, London, 1990, 3-16.

ical advancement and the so-called moral progress, and further, that numerous aspects of modern technology create dangers to culture and human life[6].

This situation forces us to re-examine what progress means, and what it should mean in contemporary science. Now we will concentrate on its particular meaning in geography.

In past Polish geography the notion of progress did not appear too frequently, particularly not until the Marxist pseudo-scientists stopped criticizing concepts which did not conform with their ideology. Lately, the concept of progress has been restored in the well-known work by Z. Chojnicki. The book was to summarize the former output of, and show directions for further works in Polish geography. He argues that, for Polish geographers " ...the most important thing in future serious research will be the progress in methods which could help us to gain factual data, and (...) introducing the rigorous techniques of field observation and remote sensing methods "[7].

The above quotation allows us to understand clearly the way the main Polish methodologist conceive progress. He emphasizes particularly its two necessary conditions : firstly, it is to be the quantity and quality of factual data, and secondly — the ways of reprocessing and utilizing them, that is to say, methods. In short, according to Chojnicki, Polish geography suffers from a growing deficit of facts and methods. Chojnicki mentioned also the need of " theoretical progress " but he failed to explain what that means. He emphasized the importance of factual data and methods in spite of the fact, that we have just experienced the " quantitative revolution ", that is to say, the period when the number of publications composed of new experimental data has been growing exponentially. The true problem has become not a shortage, but an excess of produced information. It should be noted, that only a small proportion of this scientific output has found any application. Most of the so-called " material " publications have not been utilized at all. Many have only been read by students of the article's author.

This is a meaningful situation which should not be overlooked. On the one side we are experiencing a lack of new ideas which could allow us to utilize, or even to grasp the huge amounts of the available information. At the same time however, what is most at a premium in Polish geography is the quantity of new facts and methods. In searching for, and over-examining the details, we are loosing sight of the big picture.

Polish geographers, who would like to get scientific promotions have to enlarge their output, that is to say, the number of contributions to scientific journals. Not all contributions are considered to be of the same significance.

6. W.L. McBride, " The Progress of Technology and the Philosophical Myth of Progress ", *op. cit.*, 35-38.

7. Z. Chojnicki (ed.), *Podstawowe problemy metodologiczne rozwoju polskiej geografii (Methodological problems of the development of Polish geography)*, Poznań, 1991, 376.

Only those are thought to be " truly scientific ", which can be labelled by the governmental Committee for Scientific Research as " material " articles. These are mainly elaborations of empirical data, and descriptions and applications of new methods. Discussions and controversial papers, including works in the field of the history of geographical thought, are considered to be relatively unimportant. As *belles lettres* devoid of any scientific value, they are continually ignored by the Committee as well as by the universities, mainly because they can not be contained into the frames of narrow specialization that have developed. Those scientists who decide to leave the comfortable niche of analytical experimentalism, and to enter the rugged way of historical, humanistic, or interdisciplinary investigations, will probably annihilate their chance for speedy promotion. According to Chojnicki's wishes, the main measure of scientific progress in Polish geography is the number of the produced experimental data and methods. The hundreds of minute details that are " scraped out " of the surface of economy or nature seem to be of a greater value than a wide gale of creative thought, which could impair the deep-rooted scientist paradigm.

Polish geography is continually steeped in positivist prejudice which requires searching for new empirical facts and to reprocess them with the use of newer and newer means and methods. For a geographer the greatest difficulty nowadays is not the question of how to resolve existing problems arising from the current state of knowledge. Instead, the majority of geographers at universities cudgel their brains over the question of how to find any special topic which has not yet been explored, and often more importantly, which topics will have a chance for publication. That is why there have appeared hundreds of highly sophisticated, or even curious subjects of so-called scientific works. It is an example of a lack of authenticity of scientific research, and it calls into question the very sense of continuation of studies. There appears a glaring disproportion between the number of scientists, the costs of research, and the value of significant output. The scores of insignificant research gain power, and the banks of information swell. Meanwhile, there grows the shortage of creative thought[8].

Analytical experimentalism aimed at production of new factual data is in fact the main direction in Polish geography. This occurs in spite of the proclaimed assertion, that creative potential of the positivist paradigm has just been exhausted. In fact, instead of new data and methods we really need new concepts and approaches, which could help us to grasp available information and to orient ourselves on the boundless waters of facts which are produced everyday by the newest generation of electronic machinery. Unfortunately, the main measure of the importance of particular academic centers are not contro-

---

8. See for instance W. Sedlak, *W pogoni za nieznanym (In the Pursuit of the Unknown)*, Lublin, 1990.

versial ideas but the quantity of researchers, grants, the available technological and financial potential which produce more and more megabytes in the form of publications necessary for no one.

To recapitulate we can ascertain that the progress in geography (and probably in all the positivist sciences) possesses the following characteristics :

1. multidirectional, quantitative expansion in amounts of factual data, methods, staff, equipments, and grants.

2. This leads to distraction of scientific aims and efforts, disintegration in the structure of science due to the narrowing of fields of interest.

3. This, in turn, leads to the appearance of numerous institutions, commissions, journals, sub-fields, each with own esoteric seminars, languages, and an exponential growth in the number of publications.

4. The competition between institutions, sub-fields and particular scientists for more financial aid ultimately then leads to the loss of the economic principle. The competition nowadays between scientists and institutions for more subsidies reminds one of the fight between politicians and parties for power and influence, in which personal connections are most important.

Competing institutions often needlessly spend money independently for the same aims, having no knowledge about the research being conducted on the other side of the hedge. Our corrupted political system allows institutions to waste huge amounts of money. Even in such small academic center like Kielce, thousands of dollars are squandered each year. For example, the ill-famed Holy Cross Monitoring Station Network, which has just showed to be practically useless because of faulty location and incompatible installations. It is because the distribution of subsidies depends neither on the creative potential nor the real needs, but instead on the number of staff, publications, laboratories, and most of all, on the personal connections and political correctness of the problem formulation (e.g. environmental and social sensitiveness). If one wants to gain a grant from the Committee of Scientific Research, it is not necessary to formulate an original idea. The most important condition is however preparing a research project which could make the committee sure that the author is an environmental protection activist or that he fights for so called social justice.

It seems to be evident that progress, which generally is perceived to be something excessively good, in geography is rather a source of difficulties. This is due to the following :

1) annoying information overload ;

2) great numbers of insignificant publications produced by pseudo-scientists devoid of any creative enlightenment ;

3) excessive specialization and disintegration ; and

4) unreasonable distribution of funds.

Paradoxically, progress in geography appears to be the very reason for its loss of coherence and identity, which could lead directly to its downfall. As such it no longer should be our objective.

It must be emphasized, that geography is not the only field in which progress should not be understood as the goal. In theoretical biology and medicine, progressive changes are synonymous with pathological processes (such as the uncontrolled growth of abnormal tissue in living things, which leads to a malignant tumour), while the correct changes are referred to as development.

To show that geographical and medical understanding of progress does not differ greatly, let us consider the medical description of neoplasm growth in the monstrous tumour called teratoma. In this skin tumour there are : " ... conglomerations of cerebral tissue which are not organized in the functional structure of a brain, many teeth, some of them fixed in alveolus sockets but without jaws and muscles which could move them. There are pieces of organs, strands of muscular tissue which can even contract, perspiratory and sebaceous glands which secrete great amounts of perspiration and tallow. Skin cells which produce skeins of long hair... That monstrous conglomerate does not however create any integrated organic system "[9].

In contemporary geography, like in teratoma, there are many different components which are chaotically distributed. Sometimes they make an impression of an organic structure, but unfortunately it shows to be an illusion. In both cases, geography and medicine, progressive changes do not lead to a functional whole. Modern and progressive geography, like teratoma, can be characterized by uncontrolled, specialized growth, sometimes showing the appearance of functional organization. This is not unlike those teeth in sockets and fingernails without fingers which do make cancerous growth progressive, but do not lead to functional organization. So instead of striving for progress in geography we should try to develop geographical concepts which promote functional organization. This is because the development in biological (and philosophical) terms denotes correct changes. Contrary to progress, development in biology possesses following peculiarities :

1) irreversible qualitative changes with a fixed direction which lead to

2) diversification and creation of integrative factors, increase in complexity and integration of structures (anti-entropy tendency) tantamount with the reaching of the higher level of organization (in cybernetic language - increase in information capacity) ;

9. I refer myself to the paper by Piotr Lenartowicz discussed during the mentioned Lublin conference in 1992 and published in the mentioned volume edited by S. Kyć, *Crisis of the progress concept — an ecological dimension, op. cit.* The quotation is on page 182. See also the book by the same author on *Elementy filozofii zjawiska biologicznego (Philosophy of biological phenomenon)*, Kraków, 1984.

3) this diversification and integration is the source and the necessary condition of energetic and material economy in construction of hierarchic structures[10].

If geography will grow further by way of progress alone, its future may be that of an organism, which happened to fall victim to cancer.

10. Compare Z. Cackowski, " Ruch, rozwój, postęp " (Movement, development, progress), *Filozofia a nauka - zarys encyklopedyczny (Philosophy and science : an encyclopaedic outline)*, Wrocław, 1987, 606-620. See also *Mały słownik terminów i pojęć filozoficznych (A Dictionary of philosophical terms and notions)*, in A. Podsiad, Z. Więckowski (eds), Warszawa, 1983, 343.

PART THREE

SCIENCE AND MUSIC

Aesthetic and Technological Aspects in Berio's *Thema* (*Omaggio a Joyce*)

Nicola SCALDAFERRI

The *Studio di Fonologia Musicale* of Milan was founded by Luciano Berio and Bruno Maderna in 1955, and in the first period of its activity it was a very active laboratory of technological, musical and linguistic experimentation[1]. In the *Studio* the composers worked with young scholars like Umberto Eco and Roberto Leydi, the singer Cathy Berberian and the technician Marino Zuccheri. In building the *Studio*'s technical equipment, the physicist Alfredo Lietti followed closely the suggestions of the musicians[2].

Very often there were collaborations : Berio and Maderna made some experiments together and realized with Leydi *Ritratto di città*, the first important electroacoustic work made in Milan. Moreover the Studio was attended by foreign composers like Henri Pousseur and John Cage.

One of the Studio's major achievements is *Thema* (*Omaggio a Joyce*). The composition formed the last part of a radiofonic documentary called *Omaggio a Joyce. Documenti sulla qualità onomatopeica del linguaggio poetico*. For clarity, I will call the composition only *Thema*, and the radiofonic transmission

---

1. About the history and the activity of the *Studio di Fonologia*, *cf.* L. Berio, " Studio di Fonologia Musicale ", *The Score*, 15 (1956), 83 ; G. Castelnuovo, " Lo Studio di Fonologia Musicale di Radio Milano ", *Elettronica*, 3 (1956), 106-107 ; L. Berio, " Prospettive nella Musica ", *Elettronica*, 3 (1956), 108-115 ; M. Wilkinson, " Two Months in the Studio di Fonologia ", *The Score*, 22 (1958), 41-48 ; L. Berio, " Eroismo elettronico ", *Nuova Rivista Musicale Italiana*, 4 (1972), 663-665 ; P. Santi, " La nascita dello Studio di Fonologia Musicale di Milano ", *Musica/Realtà*, 14 (1984), 167-188 ; A. Vidolin, " Avevamo nove oscillatori ", *I Quaderni della Civica Scuola di Musica*, 21-22 (1992), 13-22 ; N. Scaldaferri, " Documenti inediti sullo Studio di Fonologia Musicale della Rai di Milano ", *Musica/Realtà*, 40 (1994), 151-166 ; N. Scaldaferri, *Musica nel laboratorio elettroacustico*, Lucca, 1997.

2. Articles of Alfredo Lietti : " Soppressore di disturbi a selezione d'ampiezza ", *Elettronica*, 5 (1955), 1-3 ; " Gli impianti tecnici dello Studio di Fonologia Musicale di Radio Milano ", *Elettronica*, 3 (1956), 116-122 ; " Évolution des moyens techniques de la musique électronique ", *Revue belge de Musicologie*, 13 (1959), 40-43 ; " La scomposizione analitica del suono ", in H. Pousseur (ed.), *La musica elettronica*, Milano, 1976, 69-78 ; " I fenomeni acustici aleatori nella musica elettronica ", *loc. cit.*, 78-86.

Omaggio a Joyce[3]. In this paper I will focus on some particular aspects of the technical realization of *Thema*. However it is important to remember the cultural context in which it was born.

Thema is the meeting point of different factors. Berio's diverse interests in the writings of James Joyce, the English language, the human voice (in both the natural and the manipulated state) originated from his relationship with the singer Cathy Berberian, but were also nourished by other people he met at the Studio. One of them was Umberto Eco, who in those years was working at RAI in Milan, and who was also studying Joyce. Berio, Eco, and Berberian with the collaboration of Roberto Leydi and other people started working on the radiophonic documentary *Omaggio a Joyce* at the beginning of 1958[4]. This documentary started with a survey of the onomatopoeic sounds in various cultures, and then it focused its attention on the beginning of the XI chapter (" Sirens ") of *Ulysses* by James Joyce. The choice of the text was due, of course, to the onomatopoeic words, but also to the musical references. Every chapter of *Ulysses* is symbolized by a human organ and is dedicated to an Art. The XI chapter is symbolized by the ear and is dedicated to Music ; moreover in this chapter Joyce use a narrative technique similar to the structure of the fugue[5]. Berio and Eco used the first part of the ouverture of this chapter, ending with the phoneme /s/ ; as we will see, this sound will be very important during the composition.

Berio wrote about this research on Joyce's text in one of his most important papers, *Poesia e musica-un'esperienza*, finished in July 1958 and published in 1959 on the periodical *Incontri Musicali*[6]. In this paper Berio explained step by step the manipulations of the text made in the documentary[7]. Among the different manipulations, one is unique : the translation of the text in French and Italian and the mixing of the three languages. This manipulation is the most important moment of the documentary. Different languages and words are

3. This paper is based on the following sources on Tape :

Omaggio a Joyce. Documenti sulla qualità onomatopeica del linguaggio poetico, a cura di Luciano Berio e Umberto Eco, Milano, Rai, 1959 (personal copy of Roberto Leydi) ; *Thema*, Source Tapes-DAT transfers from original tapes (A original reading ; B track 1 ; C track 2 ; D track 3 ; E track 4 ; F mixed), Firenze, Tempo Reale, 23/5/1990.

Other musical analysis of *Thema* : F. Delalande, " L'*Omaggio a Joyce* de Luciano Berio ", *Musique en jeu*, 15 (1974), 45-54 ; N. Dressen, " Sprache und Musik als Bereiche eines Klangkontinuums : *Thema-Omaggio a Joyce* ", *Sprache und Musik bei Luciano Berio*, in G. Bosse (ed.), 1982, 38-57 ; I. Stoianova, " *Thema (Omaggio a Joyce)* ", *Luciano Berio. Chemins en musique*, Paris, 1985, 148-157 ; P. Zavagna, " *Thema (Omaggio a Joyce)* di Luciano Berio : un'analisi ", *I Quaderni della Civica Scuola di Musica*, 11-12 (1992), 58-64.

4. *Cf.* U. Eco, *Opera Aperta*, 3th edition, Milano, 1995, V-VI.

5. *Cf.* S. Gilbert, *James Joyce's Ulysses*, 2nd edition, London, 1952, 247-248.

6. L. Berio, " Poesia e musica - un'esperienza ", *Incontri Musicali*, 3 (1959), 98-111 ; reprinted in A. Lietti, " La scomposizione analitica del suono ", *op. cit.*, 124-135.

7. The readers of the texts used during the manipulations are : Cathy Berberian (original English text), Umberto Eco and Marise Flach (French translation), Nicoletta Rizzi, Furio Colombo and Ruggero De Daninos (Italian translation).

mixed regardless of the original structure of the text and of their meaning ; Berio's wish was to pay special attention to particular relations between the sounds.

He wrote : " With the meeting of three different languages, we are immediately stimulated to take in, above all, the purely sonorous aspects of the mixture, more than to pursue the various linguistic meanings ; in the presence of different messages spoken simultaneously, we can understand only one, while the others automatically become musical complements of a proper polyphonic plot " (see fig. 1)[8].

FIGURE 1.

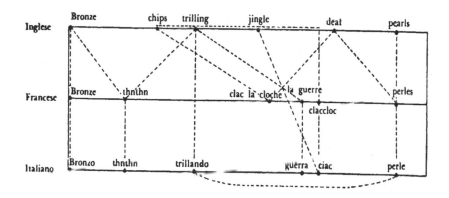

Sound relations between the three languages.

The tapes with vocal elements made during the documentary are the base for *Thema*. From all the tapes, Berio picked out phonetic elements without original significant references and made them the basic elements for the composition ; he chose the original English text only, read by Berberian, and some characteristic elements from other languages (like the Italian /r/ and the French *petites ripes*[9]).

Thema is an electroacoustic work composed in four tracks with vocal material only[10]. The piece begins with Berberian's reading of Joyce's texts for 2'20" ; afterwards Berio's electroacoustic manipulation follows, which lasts 6'22". All elements of *Thema* have their place between two extreme poles. At one pole there are vocal sounds without manipulation, clearly recognisable ; at

---

8. *Cf.* L. Berio, " Poesia e musica - un'esperienza ", *op. cit.*, 104.

9. L. Berio, " Poesia e musica - un'esperienza ", *op. cit.*, 107.

10. The first version was performed in Napoli (11-6-1958) during a concert of the season *Incontri Musicali*. The composition was restored two times : in 1990 by Paolo Zavagna on analogic support, and in 1995 on digital support by Nicola Bernardini at Tempo Reale, Firenze.

the other extreme there are manipulated sounds hardly recognisable as vocal. Between these two poles there are different levels of manipulation, with sounds that we can more or less recognise. The most interesting aspects of *Thema* are the contrasts between and the mixtures of recognisable and non-recognisable vocal sounds.

During the composition, Berio used a limited number of technical operations : filtering, echo, modulations (especially dynamic modulation), change of speed, and, of course, mixing of different tape fragments. To understand the real structure of the piece, I analyzed its first 52 seconds section, studying separately four tracks[11]. In this first section, we have six sound categories, indicated with letters from A to F[12].

A) Vocal elements (phonemes or words) slightly or not at all manipulated (for example echo effect) but in any case recognisable as vocal.

B) Events obtained with horizontal manipulation of vocal fragments ; they are recognisable as vocal, but not naturally produced by the human vocal organ. In the first section it is the famous sound /blblblblbl/ ending with the word *blooming* (45"-52").

C) Events obtained with vertical manipulation, such as the overlapping of several vocal elements. The most remarkable examples are the word-chords described by Berio in his paper. In fact Berio groups in chords words with the same vowels and classifies them according to linguistic criteria.

Figure 2 shows these word-chords[13]. In the analyzed section the chords are made with the words *sail* and *veil*.

D) Filtering. Usually Berio favors subtractive synthesis. In *Thema*, there is not white noise for filtering ; but Berio uses the /s/ phoneme like a white noise and filtered it.

E) /s/ phoneme. It is the most evident vocal sound of the entire composition and plays an important role in the formal structure ; for example it appears in the last section of Berberian's reading as well as in the beginning and in the last section of the electroacoustic manipulation.

F) Silence. To create silence the composers usually used a white tape ; however in particular cases, it was also possible to use a segment of tape with background noise. This prevents the experience of total silence, especially in pauses inside the composition.

11. The four separated tracks were given to me by Paolo Zavagna, with the permission of the Centro Tempo Reale, Firenze.

12. These categories are not perceptive ; they give in account the complexity of the technical manipulations of the tape. This method of classification of the elements of electroacoustic compositions is the result of discussions with Marino Zuccheri, the technician who materially realised electronic works in the *Studio di Fonologia*. Cf. N. Scaldaferri, *Musica nel laboratorio elettroacustico, op. cit.*, especially chapter IV (" Dall'evento sonoro al processo compositivo : analisi di *Notturno*, di Bruno Maderna ", 89-130).

13. *Cf.* L. Berio, " Poesia e musica-un'esperienza ", *op. cit.*, 105-106.

FIGURE 2.

(team)	(tip)	(tape)		(time)
steelyringing	chips	fade	sail	by
Peep	picking	awave	vell	Indolores
leave	tink	Waves		cried
sweetheart	pity			dyng
feel	jingle			bright
	listen			I
	liszt			thigh
	hiss			night
				spiked
				winding
				silent

(never)	(ever)	(tap)	(far)	(ton)	(talk)
breast	heard	satin	bronze	thnthnthn	more
never	answer	rang	stars	not	call
deaf	have	clacked	garter	love	morn
sad	pearls	smack	throb	tup	warm
when		pat	ah	come	war
		pad	Martha		all
		plash	far		lost
					saw
					call
					roar

(town)	(tone)	(took)	(tool)	(few)
rebound	gold	look	blue	blew
now	note	could	bloom	fluted
avowal	rose	full	who's	lure
jaunted	chords	good	bloo	alluring
	horn	took	bloomed	
	hawhorn		moonlit	
	so lonely		blooming	
	cold			

Word-chords.

Every track of *Thema* is the arrival point of complex editing operations and has a very contrapuntal structure ; we can find in all tracks the different elements mentioned above, mixed in a homogeneous way.

NICOLA SCALDAFERRI

FIGURE 3.

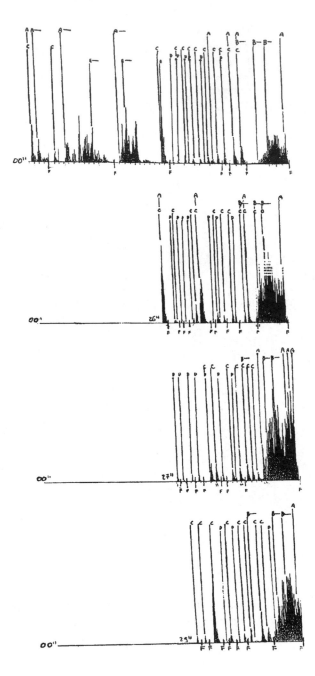

Thema, tracks 1-4, 00"-52".

In figure 3, you see the separated tracks of the first 52 seconds of *Thema* with identification of the different elements (A...F). The vertical dimension in the example is dynamics[14]. In the first 25 seconds, only the first track is playing, elements A, C and E. From 26" to the end of the section, there is a complex construction in which we can see two different moments. From 26" to 42", there are elements C, D and a few elements of A in a very close construction. After 42" there dominates the B element (made with the manipulation of the consonants /b/ and /l/).

The final mixing of different tracks doesn't make new elements but amplifies the elements of the first track and distributes the sound in the space.

In the *Studio di Fonologia*, for the creation of complex materials, composers (especially Berio and Maderna) didn't start from a rigorous plan ; however the absence of a plan (like a serial plan) is not the same as a free improvisation. Composers usually follow a path articulated in two stages :

1) Accurate choice of basis elements

2) Repetitions and variations of elements with different technological manipulations.

FIGURE 4.

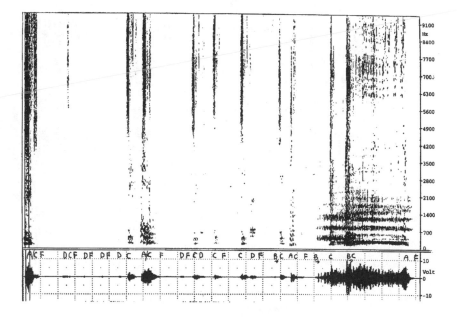

Sonogram, *Thema*, Track 2, 26"-52".

14. Figure 3 shows the original sound analysis made by Paolo Zavagna. Four tracks are superimposed. In these analysis I indicated the different elements A...F.

From the huge quantity of vocal sounds of *Omaggio a Joyce*, Berio chose only a few elements ; he composed *Thema* by manipulating these elements in the way suggested by the technical equipment. With some detailed sonograms we can understand more easily some manipulations. Now let us focus on the second track (fig. 4). This is the sonogram of the sound's elements of the second track (26"-52") in which are indicated the elements as explained above[15]. As I mentioned before, in the first passage (26"-42") we have several elements A, C, and D, very close to each other. In the second passage (43"-52") we have the element B which is very prominent. Element C occurs at different times, what is even more obvious in the detail in fig. 5.

FIGURE 5.

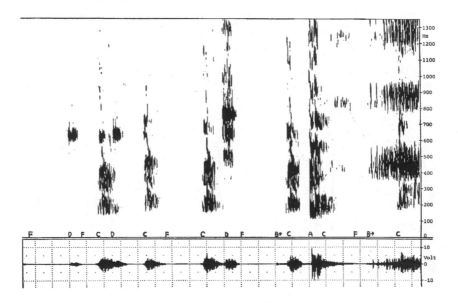

Sonogram, *Thema*, Track 2, 37"-48", detail elements C.

In this passage C element (made using the words *veil* and *sail*) appears six times. The same tape fragment is used as basis for the different technical manipulations, resulting in the progressive change of the chord. Each chord is recorded with a different dynamic and, most importantly, with a different tape speed. The change of speed, used very often in electroacustical manipulation, results in differences in the pitch and length of the sound. So we have six different versions of the same chord C.

15. The sonograms were made at the Informatic Laboratory of the Dipartimento di Musica e Spettacolo - Università di Bologna, with the assistance of Fabio Regazzi, using *Sound Scope* in Macintosh.

Another example of the change of speed (but with a very different function) is the B element in the second track, the famous *blblblblblblooming*. It is based on many fragments of tape in which phonemes /b/ and /l/ are recorded. The tape is edited a second time with change of speed ; the two excerpts are then superimposed (see fig. 6).

FIGURE 6.

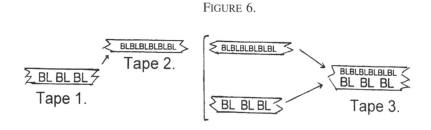

Element B, editing processus.

In *Thema*, Berio used a limited number of elements (such as the C and B elements) and manipulated them many times in different ways. This fact is rather evident if you listen carefully to the composition, and is also confirmed by Paolo Zavagna, who studied all *Thema*'s sounds (verifying results with Marino Zuccheri) for remaking the synchronisation of tracks during the restoration of the work in 1990.

Berio's compositional process in electroacoustic works is very analogous to his compositional process for traditional instruments in these years : few elements repeated in always different variations. If we consider other compositions of the same period, such as Stockhausen's *Gesang der Jünglinge*, the singularity of *Thema* is that the elements are only vocal sounds and the attention is focused in non-semantic aspects. It opened new perspectives in Berio's activity. The idea to work with non-semantic vocal sounds becomes increasingly important in Berio's later compositions, whether using tape as in *Visage*, or using natural voice, as in *Sequenza III*.

I would thank particularly Pascal Decroupet, Emily Snyder Laugesen, Fabio Regazzi and Paolo Zavagna. Responsibility for any inaccuracy or infelicity is, of course, entirely mine.

" JEU DE MIROIR *SUS UNE FONTAYNE* "

Roberto DOATI

DE MUSA ET MUSICA

When I received the invitation to realize a new piece with electronics at the *Centre de Recherches et Formation Musicales de Wallonie*, it has been natural for me, living in Padova, to think to Johannes Ciconia (1340-1411). Not only because the great composer and theorist from Liège lived his lest years in Padova, but also for the deep interaction between science and music in his life and work. He shared the same feelings with Prosdocimus de Beldemandis, mathematician, astronomer, doctor and writer of musical treatises — a truly representative of the medieval *quadrivium* conception — who was working in Padova in the first decade of 15th century.

In his treatise *De Proportionibus*[1] Ciconia writes : *Ille proprie musicus est qui musicam habet speculativam*. This medieval Latin can be translated " The true composer is who makes a music which arises from [*or* which produces] a reflection ". Reflection to be intended not only as thought, speculation, but also as formal reflecting. As a master of fact Ciconia is initiator of a big tradition which through musical structure based on canons will give rise to some of the greatest intellectual works in Western culture[2]. Important figures of this tradition are composers from French and Flemish countries such as Binchois (Mons, *ca.* 1400), Dufay (?, *ca.* 1400), Ockeghem (Dendermonde ?, *ca.* 1428), Desprez (Beaurevoir, *ca.* 1440), Obrecht (Bergen-op-Zoom, *ca.* 1450).

My choice to work on the *virelai*[3] *Sus une fontayne* is already a first reflection game. To show my veneration for the *Maestro* from Liège I choose this work because through several quotations Ciconia himself renders homage to a great musician and theorist he admired : Philipoctus de Caserta. But most important of all this *virelai* is an impressive example of what is defined *ars subtilior,* an art of proportions from big complexity.

---

1. J. Ciconia, " Nova musica. Liber tertius ", *Thesaurus Musicarum Latinarum*, vol. 9. *Greek and Latin Music Theory*, 1993.
2. Think to the influence this important tradition had on composers such as Bach and Webern.
3. Poetic and musical form of the troubadour in *oïl* language.

The term *subtilitas* appears in fourteenth century musical treatises to indicate a sort of rhythmic distortion of the melodic line. The purpose is to augment the independence of this line in front of the harmonic texture[4]. The so distorted melody — usually the *cantus* — fits into a polyphonic context where each part *(cantus, tenor, contratenor)* follows its own " grammar " which is more or less peculiar and individual. The *tenor,* straight and formal, the *contratenor,* virtuosistic and fragmentary counter-part, sometimes canonical or mathematical, and the *cantus,* " distorted " and winding. The achievement is a complex — but still comprehensible — play between three totally autonomous parts, which give rise to a diffuse rarefied atmosphere. The *subtilitas* begins in France — particularly at the Avignon court — and soon reaches the pro-French courts in Italy (Pavia, Milano, Modena).

In Italy the *subtilitas* becomes *subtilior*[5] — term used for the first time by Prosdocimus. The complexity of the weaving counterpoint is increased and the *subtilitas* undergoes a dramatization. The transgression is subjected to the Italian peculiar need to represent.

ACOUSTIC TREATMENTS

Chance, intuition and my Italian culture made me dramatize the single parts of the *virelai,* shaping them on the temperament of the musicians I have worked with. I am always interested in writing music which in some way reflects the psychological and interpretative character of the musicians I am working with. In this case the " grammar " used for each part recalls both the voices character of the *ars subtilior* and personality of Izumi Okubo *(cantus,* violin), Catherine Binard *(contratenor,* flutes), Jean-Pierre Peuvion *(tenor,* clarinets). The instrumental score is a " corrupted " transcription of the *virelai* three voices for violin, flute and clarinet. Each note from Ciconia's work is " transformed " on the score following 5 treatments classes (fig. 1).

FIGURE 1.

Class	Group	Examples
A	Time transformation	flutter-tonguing, *jeté*, trills, *acciaccature*
B	Pitch transformation	mikrotones, *glissandi*, transpositions
C	Dynamics transformation	flutter-tonguing, *tremolo* & its variations
D1	Timbral harmonic transformation	timbral trills, harmonics
D2	Timbral inharmonic transformation	multiphonics, *pizzicato*, keystrokes
E	Interpolation	transitions between different manners

Classes of acoustic treatments for *Felix Regula.*

4. Anonymous treatise, *Tractatus figurarum*, 1989.
5. U. Günther, " Das Ende der Ars Nova ", *Die Musik-Forschung*, 16 (1963), 105-120.

In figure 2 we can compare the modem notation[6] of *virelai* first measures (a) and its treatments (b) through the application of classes in fig. 1. The association between notes and classes is based on a musicological and sometimes visionary analysis of *Sus une fontayne*. Comparing a performance of the original *virelai*[7] the metronome has been slowed clown about 5 times (crotchet = 35). For formal reasons, besides the rhythmic and timbral articulation, I modified — only twice — the original interplay of the parts. In the example of figure 2, I doubled the durations on the *cantus* measures, and in another place I subtracted 6 measures to the *contratenor* part. The fact that these changes do not alter anyway the "architecture" of the work is a demonstration that in fourteenth century compositions were often conceived as a sort of assembly kit. This is one of the reasons why so many contemporary composers fell attracted by Middle Ages music.

Each instrumental part has been recorded in studio — musical assistant Jean-Marc Sullon — apart from the other to maintain the same character of detachment of the *virelai*.

FIGURE 2a.

First measures *Sus une fontayne* ;

6. I. Bent (ed.), "Johannes Ciconia", *Polyphonic Music of the 14th Century*, Monaco, 1978-1990, 170-174.

7. Ensembles "Alla francesca" and "Alta", *Johannes Ciconia*, CD OPS 30-101, Paris, 1994.

FIGURE 2b.

First measures *IV Felix Regula.*

ELECTRONIC TREATMENTS

The formal relationship between each instrument and its own electronic transformation have been thus determined as shown in figure 3.

FIGURE 3.

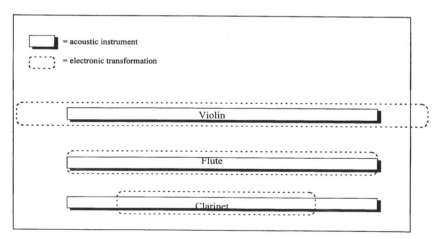

Formal schema for *IV Felix Regula.*

All the violin sounds are transformed and symmetrically distributed, but time stretched, around the middle of the violin part. So the first measures played by the instrument will be heard, electronically transformed, 50 seconds before their real production, while the lest electronic sounds are placed 50 seconds after their acoustic originals. The electronic transformation of flute is always simultaneous with the instrumental part, in a sort of live electronics simulation. The clarinet sounds, finally, are transformed, time compressed and symmetrically distributed around the middle of the instrumental part. The first measures played by the instrument will be heard electronically transformed after 40 seconds, while the lest electronic sounds will appear several seconds before their acoustic originals.

Then I choose the digital tools from those available at the *Centre de Recherches et Formation Musicales de Wallonie* in Liège. All the treatments of the instrumental sounds were clone on a Power Mac with Audiosculpt v. 1.2bl (IRCAM), Lemur Pro 4.0.1 (CERL Sound Group, University of Illinois) and MAX (Opcode) software. In this case too I built a timbre space to organize the electronic transformation (fig. 4). It is composed by 15 different processes — some of them with one or more variants — associated to the 5 acoustic treatments classes.

FIGURE 4.

Class	Sound production	Electronic processes
A	flutter-tonguing, *tremolo*	Cross synthesis with timbral trills
A	Timbral trills	*Vibrato* superposing
A	Timbral trills, *jeté, pizzicato*	Cross synthesis with multiphonics
A	flutter-tonguing , timbral trills, *jeté*	Time stretching of grains
B	Normal	Formantic filtering and microtonal textures
B	Normal	Spectrum transposition
B	Normal (low and high pitches)	Harmonizing
B	Mikrotones, random fragments	Dynamic time stretching
C	flutter-tonguing, *tremolo* and their variations	Cross synthesis with timbral trills
D1	Whistle tones	Spectrum stretching and compressing
D2	Multiphonics	Stocastic part extraction
D2	Slap, keystrokes, *pizzicato*	Time stretching and transposition
D2	Breathing, bow dragging	Filtering with analysis file
D2	Multiphonics	Dynamic filtering (*glissando*)
E	Transition between different productions	Granular morphing

Classes of electronic treatments for *Felix Regula*.

We will see now an example of transformation on violin sounds. Through the graphic filtering allowed by Audiosculpt on the sonogram of a *col legno* produced sound — very noisy — armonics are deleted (white rectangles in figure 5a). Then, with the same filtering tool, noisy regions of the spectrum are emphasized (+20 dB). See darker areas in figure 5b.

FIGURE 5a.

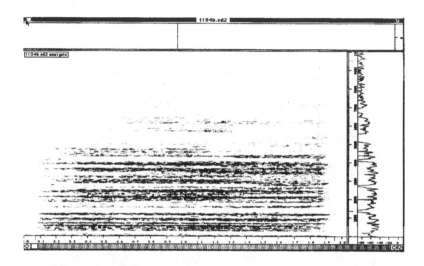

Deleting armonics from a violin sound (*col legno*) ;

FIGURE 5b.

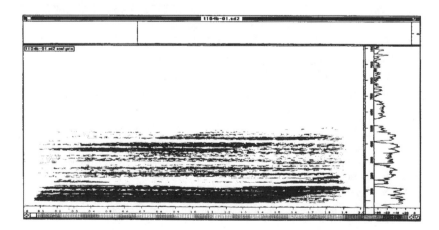

Emphasizing noisy regions.

5 HAPPY RULES

What results from all the instrumental and electronic parts is a 8 tracks tape with mono recordings of the instrument and " electronic clarinet ", and stereophonic recordings of violin and flute electronic transformations. This tape can be freely interpreted by a performer as concern tracks (minimum 3 tracks), dynamics and spatialization. During the mixing process I realized that each instruments with its own electronics could have independent life, which is true also for the single voices in Ciconia's *virelai*. So I decided to name *I Felix Regula* the clarinet and tape version, *II Felix Regula* the flute and tape version *III Felix Regula* the violin and tape version. *V Felix Regula* is a really different version. Because of the scoring which asks for the live instruments, *IV Felix Regula* tape and live electronics, but also from a formal point of view. Although the sound materials are quite the same, the structural relationships between instrument and electronics are changed, again in a mirroring game (fig. 6).

FIGURE 6.

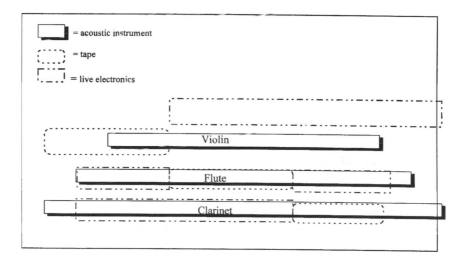

Formal schema for *V Felix Regula.*

Live electronics is realized with the Next based IRCAM workstation - 2 DSP. Tape and live electronics are both used as safety margin during the performance and for a processing reduction reason. In fact I use only two of the four inputs available for live electronics, the third instrument, in turn, being supported by a tape part. Fig. 7 shows general schema of the MAX patches used.

FIGURE 7.

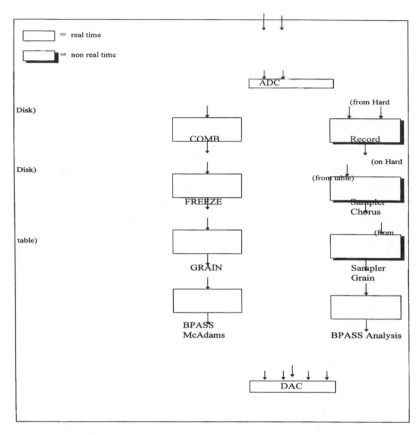

General schema of MAX patches used for *V Felix Regula*.

ADC = analog to digital conversion

COMB = stochastic part extraction

FREEZE = loop from a fragment, with possible transposition (4 modules)

GRAIN = random granulation (4 modules)

BPASS McAdams = band pass filtering with spectrum stretching and compressing (11 modules)

BPASS Analysis = filtering with amplitude/frequency determined by FFT analysis file (10 modules)

Record = table writing

Sampler Chorus = table reading for transposition and chorus (3 modules)

Sampler Grain = table reading for granulation (4 modules)

DAC = digital to analogue conversion

As concern the performance of the live electronics part, it was decided to write all the patches parameters in a list whose reading is controlled by a MIDI pedal activated by one of the performers. This allows a precision in the transforming algorithms comparable to that used to realize the tape. Dynamics, reverberation and spatialization are to be programmed each time according the acoustic characteristics of the concert space.

Re-Synthesis of Analogue Electronic Music Compositions by Computer : A Teaching Experience

Alvise Vidolin

Electronic Against Traditional Music

Past music passed on to us by score (and not by sound), by traditional musical instruments, and by performer activity who mediates music with social mutation (function, space, aesthetic, etc.). Particularly, each performance is a re-creation of the compositional thought and teaching activity is based on oral transmission of musical language and performance practice.

On the other side, electronic music passed on to us only by sound : without any score, notation language, and standard musical instruments. An electronic music composition is a sound document entirely determined by the composer, and is very rare to have different musical interpretations.

The sound document contains the composer's musical thought, sonic creation, and performance. In other words, contains both compositional and performance work, both done by composer himself.

We can end by saying that electronic music differs from acoustic music in composition, performance, and, as we will see soon, in teaching.

Electronic Music Teaching

Musical analysis is essential for teaching any kind of music and it is important also for teaching electronic music. Because only few pieces have a score, and also for them, new analysis methods are needed.

We can distinguish three types of electronic music analysis. Perceptual analysis studies form and macro-temporal aspects, segmentation and internal dynamics, and compositional meanings. Sonological analysis studies sound morphology and micro-temporal aspects. Document analysis is based on score and composer's sketches or notes, if these exist.

I am using analysis by synthesis method for teaching electronic music because it work fairly well for sonological analysis. We recreate parts of " classical " electronic music composition and we can have instant goal verification if the copy differs from the original. Moreover the student has a good ear training for electronic sounds and for the knowledge of electronic music repertoire. Exercises can differ in degree of difficulty depending upon score presence, compositional notes, and analytical works.

COPY FROM THE ORIGINAL

When we have the sound document only, the copy from the original requires an accurate sonological analysis. If the original process is unknown, there are many ways to reach the same outcome and this can be used for teaching different solution. The copy from the original is a good teaching method because the student must make a deep analysis, design synthesis algorithms, produce the operational score, and verify the sonological result between the original and its copy. This processes can be looped many times for having a better result.

COPY FROM A SCORE

Electronic music repertoire has not many scores. Most of the scores comes from the electronic music compositions made at *Studio für Elektronische Musik des Westdeutschen Rundfunks* in Köln and printed by Universal Editions. The copy from a score is a easier task compared to the copy from the original, but it presents some philological problems. Electronic music score differs from the traditional score because a common notation language does not exist. Electronic scores are like ancient tablatures. The composer indicates technical operations and not musical (or perceptual) results ; score data often refers to specific equipment parameters and so is difficult to get hold of the technical documentation of this equipment.

RE-SYNTHESIS EXAMPLES

I'm going to present three re-synthesis examples made at the *Conservatorio* " Benedetto Marcello " in Venice as teaching work with some student of mine. Because of editorial constrain, it was not possible to include sound examples in this proceedings. I apologise the reader fort the lost of information. The examples that I am going to present, Franco Evangelisti *Incontri di fasce sonore*[1] (1957) and Gottfried Michael Koenig *Essay*[2] (1957) were done by Massimo Marchi ; Karlheinz Stockhausen *Kontakte* (1959/60) was done by

---

1. F. Evangelisti, *Studio Elettronico. Incontri di fasce sonore*, UE 12863, Wien, 1957.
2. G.M. Koenig, *Essay - composition for electronic sounds - 1957*, UE 12885, Wien, 1960.

Mario Colbacchini. All these pieces has a score so my examples seem to present only the problem of a copy from the score. But, as we will see later, the score is not exhaustive and complete, and very often is necessary to compare the copy with the original. Moreover these three pieces show three different types of score so I can present different types of problems. *Incontri di fasce sonore* is a graphic/scientific score, *Essay* is a procedural score (instructions for realisation) and *Kontakte* has two scores : the performance score[3] and the construction score[4]. Sound examples was made by computer using the synthesis program Csound[5].

INCONTRI DI FASCE SONORE

Incontri di fasce sonore is a graphic/scientific score. The *Leporello* form was designed to give both the idea of the structure and the form than the possibility of reading the music in a normal way. As shown in fig. 1, the score is horizontally divided into four parts. From the top there are two systems of lines numbered from 1-21 and 21-1 in which are signed the sound groups and their elements. After down we find the time line where is defined the duration of each element (centimetres of tape ; tape speed = 76.2 cm/s) and, at the bottom, the dynamic level (dB) of each element.

FIGURE 1

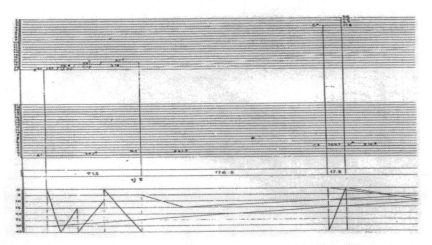

Franco Evangelisti *Incontri di fasce sonore*. Score. Universal Edition 12 863.

3. K. Stockhausen, *Kontakte - Aufführungspartitur*, UE 14246, Wien, 1960.

4. K. Stockhausen, *Kontakte - Elektronische Musik - Realisationpartitur*, UE 13678, Wien, 1960.

5. B. Vercoe, *Csound - A Manual for the Audio Processing System and Supporting Programs*, 1986-1998.

Fig. 2 compares the amplitude envelope of the first second of the original with the copy. The graph has sound amplitude on ordinate (linear scale) and time on abscissa. We can see that the amplitude envelope of the copy is closer to the symbolic lines of the score than the original.

The original is clipped and this distortion is audible also listening the sound example 1 (original) and sound example 2 (copy).

FIGURE 2.

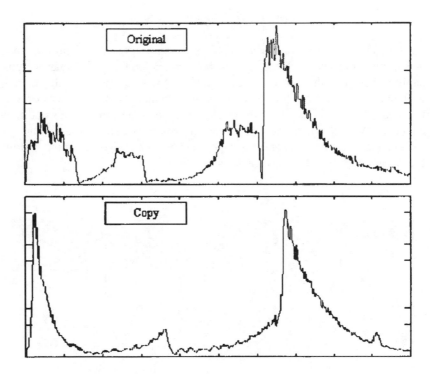

Franco Evangelisti, *Incontri di fasce sonore*.
Amplitude envelope from 0 to 1' 01".

The problem that we see in the first second of the music can be seen also in a longer frame of music.

Fig. 3 shows the amplitude envelope of the first four seconds of music. We can see that the copy follows exactly the score.

FIGURE 3.

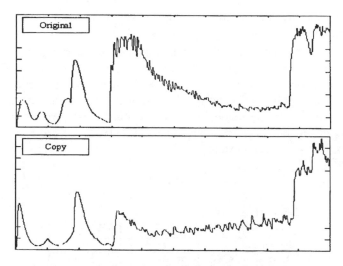

F. Evangelisti, *Incontri di fasce sonore*. Amplitude envelope from 0 to 1' 01".

FIGURE 4.

```
110          MATERIAL A                    MATERIAL A

111          Gesamtlänge: 384.7 cm.        Total length: 384.7 cm.
             Teilung in 7 Sektionen:       Division into 7 sections:
                              12.0 cm  )
                              17.9     )
                              26.9     )   Quotient:    3/2
                              40.4     )
                              60.5     )   Reihenfolge:
                              90.8     )   Sequence:    1 2 5 4 3 7 6
                             136.2     )

             Jede Sektion enthält 8 Einzel-   Each section contains 8 single
             werte:                           values:
     .1          (x+y) (6)    1.1 cm  )
                              1.2     )
                              1.3     )   Quotient:    12/11 (7)
                              1.4     )
                              1.5     )   Reihenfolge:
                              1.7     )   Sequence:    5 4 1 7 6 2 8 3
                              1.8     )
                              2.0     )
                             ─────
                             12.0 cm

     .2                       1.6 cm  )
                              1.7     )
                              1.9     )   Quotient:    11/10
                              2.1     )
                              2.3     )   Reihenfolge:
                              2.5     )   Sequence:    4 3 8 6 8 1 7 2
                              2.7     )
                              3.0     )
                             ─────
                             17.9 cm
```

Gottfied Michael Koenig, *Essay*. Score, page 13.

ESSAY

Koenig's *Essay* is a procedural score. As we see on fig. 4 the score is a list of realisation instructions.

Table 1 shows the genesis of the first part of the work : part A. Part A is made by 7 little section : each of them is a sequence of 8 elements. These elements are defined by duration, frequency and timbre. Timbre can be sinus (s), noise (n), and pulse (i) sound. Each section is transformed by 5 types of transformations : transposition, ring modulation, filtering, reverberation, and intensity modulation. The resulting sounds are mixed on three layers. Table 2 presents the list of sound examples.

TABLE 1.

Material A

Length [cm]	384.7
Sections	7
Duration Quotient	3/2
Duration Init. [cm]	12.0
Duration Sequence	1254376
Frequency curve	⊏ ⊐
Frequency ambit [Hz]	400-800
Frequency Quotient	$^8\sqrt{2}$

Frequency ambit

Ambit	1	2	3	4	5	6	7	8
Freq. range	400 436	436 476	476 519	519 566	566 617	617 673	673 734	734 800

Part A

	Start[cm]	Mat. [t]	E.D. [cm]	Mat. [t]	E.D. [cm]	Mat. [t]	E.D. [cm]	Mat. [t]	E.D. [cm]	Mat. [t]
Layer I	0	1	116.3	2	763.4	4	392.6	5	261.7	6
Layer II	290.8	3								
Layer III	2859.0	7								

Base Material A

Section		1	2	3	4	5	6	7
Elements		8	8	8	8	8	8	8
Duration	[cm]	12.0	17.9	60.5	40.4	26.9	136.2	90.8
	Init. [cm]	1.1	1.6	4.5	3.2	2.3	8.3	6.2
	Quotient	12/11	11/10	8/7	9/8	10/9	6/5	7/6
	Sequence	54176283	43865172	18532647	76318425	65287314	21643758	87421536
Frequency	Curves	⩒	/	▽	✕	△	/	‾
	Type	(x+y)		(x+y)	(x+y)	(x+y)		(x+y)
	Ambit	1-7	3-8	4-5	1-8	3-5	2-6	3-6
	Init. x [Hz]	400	476	566	400	476	436	566
	Quotient x	$\sqrt[8]{519/400}$	$\sqrt[8]{800/476}$	$\sqrt[8]{617/566}$	$\sqrt[8]{800/400}$	$\sqrt[8]{519/476}$	$\sqrt[8]{673/436}$	$\sqrt[8]{673/566}$
	Sequence x	54176283	12345678	54176283	23456789	54176283	87654321	54176283
	Init. y [Hz]	519		519	400	519		476
	Quotient y	$\sqrt[8]{734/519}$		$\sqrt[8]{566/519}$	$\sqrt[8]{800/400}$	$\sqrt[8]{617/519}$		$\sqrt[8]{566/476}$
	Sequence y	86421357		75312468	87654321	24687531		13578642
Timbre		sssssnss	ssnnsssn	nnssnnnn	snnnnnnii	nnnnniinn	niinnnii	iiniiiii

Transformations of Material A

Transformation	1	2	3	4	5	6	7
Transposition	200/50	---	6.25/50	100/50	400/50	12.5/50	25/50
Resulting lengths [cm]	96.2	384.7	3078.0	192.4	48.1	1539.0	769.5
Ring-modulation [Hz]	750	1200	5400	900	675	3000	1800
Ring-carrier type	S	S	S	S	S	S	S
Filtering [Hz]	400-800	400-800	---	400-800	400-800	400-800	400-800
Reverberation			Constant				Increasing
Intensity [dB]	0 - -40	0 - -40 - 0	Min - -20	-40 -0 -40	-40 - 0	0 -40 -0 - / -40	-30 - -5 - - / 20 - 0
Other transformations	---	---	---	---	---	---	---

G.M. Koenig, *Essay*. Part A.

TABLE 2.

Sound example	*Essay* Description	Remarks
3	Base material	
4	Transposition	1) 2 octaves up ; 2) original ; 3) 3 octaves down
5	Ring modulation	1) carrier = 750 Hz ; 2) carrier = 1200 Hz ; 3) carrier = 5400 Hz
6	Filtering	1) 400-800 Hz ; 2) 400-800 Hz
7	Reverberation	3) constant
8	Intensity modulation	1) 0 - - 40 dB ; 2) 0 - - 40 - 0 dB
9	Part A copy	
10	Part A original	

G.M. Koenig, *Essay*. Sound examples list.

KONTAKTE

Problems that we found on *Incontri di Fasce sonore* and on *Essay* are more evident on Stockhausen's *Kontakte*. As I said before, *Kontakte* has two scores : the performance score and the construction score. Fig. 5 shows the section X : pages 19-20 of the performance score in which electronics are notated in a free form.

FIGURE 5.

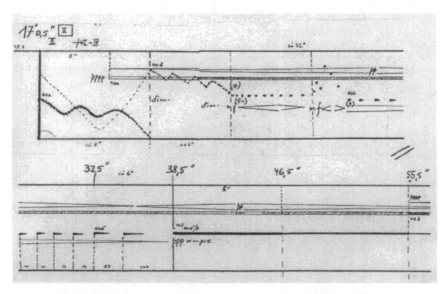

K. Stockhausen *Kontakte*. Performance score, pages 19-20.

Time runs on the horizontal line and frequency on the vertical line. In this score we can read also the dynamic, in musical notation, and the elements label. The Construction score describes the construction of the each element, as shown in fig. 6. It is very precise and it gives a good documentation of the work made by Stockhausen. Making the re-synthesis of the score showed on fig. 5 we found some problems. Stockhausen gives only operational and not technical description of the electronic equipment. For instance, he describes the control panel of the filter but he does not give it pulse response which is the precise measure of the filter behaviour.

The sound examples 11 and 12 present the original and the copy of this part of the score. Also in this piece the main difference is sound quality : the copy has a sound cleaner then the original.

FIGURE 6.

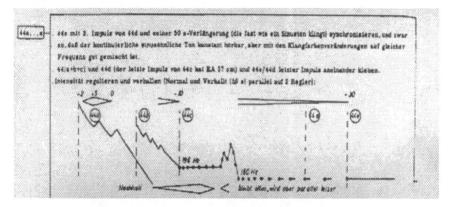

K. Stockhausen, *Kontakte*. Construction score, page 48.

CONCLUSION

Re-synthesis of analog electronic music compositions by computer is a good teaching method because the student must make a deep analysis of the composition, involving both formal and sonological aspects. If we compare the original to the computer generated copy we ear many little differences. This differences can depend on technical reason but can be also associated to the freedom of the musical interpretation.

The performance of an electronic music score can go behind the teaching exercise and become a musical performance. In this case it is difficult do not compare the new performance to the original one made by the composer, and it is natural the question : may we consider the original tape an *unicum* ?

As I said before, past music passed on to us by score and by performer activity. Why not to continue this tradition also for electronic music ?

Riflessione sui sincretismi e l'evoluzione della musica latinoamericana negli ultimi cinque secoli[1]

Carla Minelli

La musica latinoamericana è un universo estremamente variegato e vitale, frutto dell'incontro tra culture diverse e cioè quella indigena, quella africana e quella europea. Da tale incontro si è sviluppato un modo di concepire la musica vivo e spontaneo, caratterizzato :

- da un rapporto tra musica popolare e colta molto accentuato ; è proprio la prima ad essere, in molti casi, fonte di ispirazione della seconda per cui molto spesso i confini tra l'una e l'altra non risultano ben definiti ;

- da un simbolismo rituale che può essere molto forte, considerato da noi europei extramusicale e che può lasciare traccia anche là dove si è perso ogni riferimento religioso ;

- dalla capacità di adeguarsi ad imposizioni, avvenute solitamente attraverso la religione, e di assumere e trasformare, facendoli propri, mode e generi musicali provenienti da altri paesi appartenenti allo stesso continenete americano o dal Vecchio Mondo.

L'Incontro di tre culture : modi diversi di concepire la musica

Attraverso la conquista, nuovi costumi, modi di pensare e linguaggi arrivarono in America ; tra questi ultimi troviamo, per l'appunto, pure quello musicale. Con i conquistatori e poi con gli schiavi africani, giunsero non solo nuovi elementi musicali e scale e ritmi diversi da quelli autoctoni : venne rivoluzionato il concetto di musica, furono cioè portate idee molto differenti riguardanti i significati più profondi di questo linguaggio.

---

1. Questo articolo vuole essere solo una riflessione riguardante le mie esperienze sulla musica latinoamericana.

In poche parole l'Europa portò una concezione della musica molto diversa da quella amerindiana : mentre presso le antiche civiltà precolombiane o le poche culture indigene ancora oggi esistenti e rimaste intatte (ad es. quelle amazzoniche) la musica era od è una necessità quasi esclusivamente rituale, per l'europeo dall'era moderna in poi, è prevalentemente un fatto estetico. Per essere più chiari, per gli amerindi la musica era od è un linguaggio divino mentre per l'europeo è una forma di preghiera che può arricchire un rito religioso come la messa o che può avere altri scopi che sono generalmente legati all'estetica ed al divertimento.

La cultura europea venne diffusa sia attraverso i centri di potere dei conquistatori, cioè città come Città del Messico e Lima, sia attraverso i missionari. Furono soprattutto questi ultimi a penetrare nella cultura locale ; scoprirono ben presto che uno dei veicoli migliori per evangelizzare era proprio la musica in quanto radicata nella cultura amerindia[2]. Anche gli africani importati come schiavi soprattutto su gran parte delle coste occidentali e delle isole dei Caraibi, portarono una nuova cultura, una loro religione e, naturalmente, anche nuovi elementi musicali e cioè strumenti, scale e ritmi della loro terra. Il loro mondo sonoro, assai variegato, possedeva e possiede tutt'oggi un significato molto forte durante i riti religiosi ; infatti anche per queste popolazioni, così come per quelle amerindie, la musica può essere intrisa di una forza rituale estranea alla nostra religione cristiana dell'epoca moderna[3].

La cultura africana portò una sua concezione della vita che venne presto a scontrarsi con quella occidentale : non solo furono spesso condannati i riti di questi schiavi ma pure molte danze considerate dai cattolici oscene a causa del loro erotismo[4] e forse pure pericolose perchè potevano fomentare centri di potere neri non controllabili. Così ad esempio nel 1760, in Argentina furono vietate dalle autorità locali i *candombes*, feste di origine africana in cui si incoronavano un re e una regina ; sei anni più tardi lo stesso vice-rè le dichiarò indecenti e ancora una volta vennero bandite nel 1770 e nel 1790. Molte di queste feste di origine africana, che furono ripetutamente bandite, andarono perdendo nel tempo il loro significato religioso e spesso si trasformarono in feste profane eseguite durante il carnevale.

La Chiesa cattolica, nel tentativo di controllare e di evangelizzare le popolazioni di origine africana, promosse la formazione di confraternite sotto la propria supervisione. Molte divinità africane presero il nome di santi cattolici ; così ad esempio a Montevideo la confraternita dei *Congos* trasferì la devozione della divinità Calunga alla figura di San Baldassarre[5]. La pressione esercitata dalla Chiesa promosse quindi sincretismi a diversi livelli tra religioni africane

2. G. Béhague, *La musica en America Latina*, Caracas, 1983, 21-26.
3. A. Ramos, *O negro brasileiro*, Recife, 1988, 162.
4. R. Bastide, *Le Americhe nere*, Firenze, 1970, 116-127, 179.
5. I. Leymarie, *Du tango au reggae. Musiques noires d'Amérique Latine et de Caraïbes*, Paris, 1996, 224.

e il cattolicesimo che determinarono la formazione di rituali dove ad esempio la croce e i santi cattolici si sono sostituiti ad antiche divinità africane. In un rito Umbanda a cui assistii il 15 agosto 1995 a Poços de Caldas (Minas Gerais, Brasile), si festeggiava *Yemanjá*, l'*orixá* delle acque salate, sotto le vesti della Madonna e nel giorno consacrato dai cattolici alla sua ascensione al cielo.

Altro elemento che contraddistingueva in maniera marcata la musica dei tre continenti, era la scrittura : solo gli occidentali conoscevano la scrittura musicale, gli amerindi e gli africani apprendevano canti, melodie e ritmi fin da bambini, a memoria, tramandati oralmente dai propri parenti o maestri.

Mentre la scrittura musicale presso la musica colta occidentale ha permesso lo sviluppo di un linguaggio assai complesso, la sua assenza presso le altre culture ha portato ad un'evoluzione differente della musica e a coltivare caratteristiche diverse come ad esempio l'improvvisazione : chi suona seguendo memonicamente un ritmo preciso e la traccia di una melodia, ogni volta ricrea il brano introducendo elementi musicali generati sul momento.

I popoli autoctoni così come quelli di origine africana, impararono comunque velocemente sia la scrittura musicale, sia ad usare strumenti musicali importati dall'Europa. Alcuni missionari intrapresero infatti una vera e propria educazione musicale, insegnando spesso agli amerindi non solo canti nuovi strutturati su scale musicali diverse da quella autoctone, ma pure la scrittura musicale e la pratica strumentale che venivano apprese con grande facilità.

Tra gli ordini religiosi presenti in America Latina, troviamo i francescani e i gesuiti. Questi ultimi che lavorarono soprattutto in America del Sud in particolare in Bolivia, Paraguay, Argentina e Brasile, crearono scuole di musica dove indios e meticci arrivarono ad una formazione musicale sufficiente tanto da poter cantare o suonare in cori e in piccole orchestre. Presto si venne a formare una schiera di musici indigeni dei quali però riuscirono a occupare posti di rilievo solo i meticci, cioè coloro che avevano sangue indio e spagnolo. In Messico e in Perù vennero organizzate scuole speciali per figli di capi indigeni ; a Quito, ad esempio, i francescani nella seconda metà del 1500 aprirono un collegio dove i figli dei capi indigeni potevano familiarizzare con il canto gregoriano e la polifonia[6].

Anche i popoli di origine africana impararono velocemente la scrittura e a suonare strumenti europei. Ad es. a Lima, nel XVII secolo, il collegio gesuita di San Pablo aveva un orchestra formata da neri che suonavano strumenti come il clarinetto, la tromba, il flauto, la chitarra. Ancora, nella seconda metà del 1700, musicisti neri insegnavano la musica classica o suonavano il violino nell'orchestra della cattedrale di Buenos Aires[7].

---

6. G. Béhague, *La musica en America Latina*, op. cit., 22-23.
7. I. Leymarie, *Du tango au reggae. Musiques noires d'Amérique Latine et de Caraïbes*, op. cit., 211-224.

Gli occidentali cercarono quindi di dominare non solo fisicamente ma anche e soprattutto culturalmente le popolazioni amerindiane e africane senza comunque, in molti casi, riuscire a sopprimerle. Si venne così a creare un sincretismo molto particolare e diverso da regione a regione.

La cultura europea si venne a fondere con elementi africani soprattutto sulle coste dell'Uruguay e del Nord dell'Argentina, del Brasile, delle Guiane, del Venezuela, della Colombia, dell'Ecuador e del Panama oltre che nelle isole caribiche e, in una percentuale molto inferiore, sulle coste di alcuni paesi dell'America Centrale.

Mentre la foresta tropicale è rimasta, per quei pochi gruppi sopravvissuti, una roccaforte della cultura indigena, nella regione andina, dove gli spagnoli presero possesso dell'impero incaico ma dove l'altitudine rendeva la vita agevole solo ai nativi, molte sono le tradizioni amerindiane ancora vive anche se generalmente in sincretismo con elementi europei.

Spesso è comunque assai difficile discernere e riconoscere la provenienza di determinati elementi : in alcune regioni come le Antille o il Nord-Est del Brasile, sicuramente l'amerindio, per certi aspetti culturali molto vicino all'africano, ha dato un suo contributo che si è venuto poi a confondere per molti aspetti alla vita rituale del negro[8].

La religione è la chiave per comprendere una cultura : è là che possiamo incontrare ancora oggi elementi la cui origine è evidente. Un esempio possono essere le religioni afro-americane. Nel campo profano, anche se si sono create aree di predominanza africana o europea o indigena (in questo caso la musica è stata spesso privata del suo elemento rituale come è avvenuto sulle Ande), le influenze e gli incroci delle tre tradizioni sono stati molto più forti e continuano ancora oggi determinando la creazione di nuove correnti musicali.

La cultura europea ha comunque dato un'impronta unificatrice in America Latina, anche se si possono notare due grandi aree : quella lusofona, cioè il Brasile, e quella di lingua spagnola. Per essere più chiari, alcune caratteristiche tipiche dell'America di lingua spagnola, ricostruibili nell'uso di determinati strumenti musicali (ad es. l'arpa), di certi ritmi che vedono il 6/8 e il 3/4 sovrapposti e nella coreografia di certe danze, non si ritrovano in Brasile.

GLI STRUMENTI MUSICALI

Tanti e vari sono gli strumenti musicali in uso in America Latina ; parte sono di origine europea, parte africana e parte amerindia. L'incontro delle tre culture ha a volte dato vita a nuovi strumenti musicali : essi sono cioè stati adattati dai musicisti locali secondo le proprie esigenze.

---

8. R. Bastide, *Le Americhe nere*, *op. cit.*, 100-115.

Gli strumenti amerindi

Questi si trovano ovviamente soprattutto presso quei pochi gruppi rimasti intatti e che vivono principalmente nella foresta tropicale. Strumenti indigeni, anche se affiancati da strumenti europei, si trovano comunque anche in stati come il Messico, il Guatemala, il Venezuela e soprattutto nella regione andina.

Gli strumenti indigeni sono costruiti con materiale che si trova facilmente in natura : canne, zucche, legno, ossa e pelli di animali, semi e conchiglie con cui si ricavano strumenti a fiato, sonagli e maracas. L'indio sia che viva isolato (ad esempio in Amazzonia) sia che ormai da tempo sia in stretto contatto con la cultura bianca (ad es. i Quechua e gli Aymara delle Ande), si costruisce il proprio strumento. Questa è un'altra caratteristica dell' amerindio che si viene a differenziare dalla cultura occidentale.

Così sulle Ande i membri di ciascun gruppo si costruiscono, ad esempio, i propri *sikus* (flauti di pan) intonandoli tra di loro ; tali strumenti potranno quindi essere suonati solo in quel determinato gruppo. Sempre sulle Ande, ma ovviamante molto di più nella foresta tropicale, sono evidenti alcune caratteristiche rituali degli strumenti. Molti di essi conservano ancora il proprio sesso[9] : i flauti (esempi andini sono la *quena*, il *quenaco*, il *sikus*, il *rondador*, l'*antara...*) sono sempre maschili e ancora oggi, vengono suonati solo da uomini. L'origine del *sikus* è inoltre con evidenza, rituale : la sdoppiatura e complementarietà del *sikus* in *ira* (flauto femmina) e *arka* (flauto maschio), richiama infatti la cosmogonia amerindia e, all'interno di essa, la necessità e la complementarietà del maschio e della femmina evidenti in tutte le forme di vita. I *sikus* ancora oggi, fuori dai centri urbani vengono suonati da due persone differenti ; una di esse suona l'*ira*, l'altra l'*arka*; i suoni dei due strumenti sono complementari e vengono emessi in maniera alternata.

Gli strumenti musicali suonati nei rituali indigeni in uso presso quelle culture amerindie ancora intatte, sono considerati sacri e, in quanto tali, sono carichi di una simbologia intrisa di parte del cosmo sacro e animato che permette lo svolgimento del rito. I loro significati si riallacciano quindi sempre alla religione del gruppo, attraverso la cosmogonia e la mitologia. Un esempio può essere il flauto di *bamboo* denominato *upawá*, che viene usato dagli indios Xavantes (Mato Grosso, Brasile) durante il rito di iniziazione maschile al *wayá*. È uno strumento maschile, suonato solo da uomini e considerato tabù per le donne. Come mi è stato riferito da Francisco, un informatore indigeno[10], il suo suono grave e lugubre imita il ronzare dell'ape nera ed è quindi, come

---

9. Ogni strumento musicale amerindio può infatti avere carattere maschile, o femminile o ancora ermafrodita (messo in evidenza dal fatto che è suonato solo dao maschi, da femmine o da entrambi i sessi), così come tutti gli strumeti del creato fanno parte delle due grandi divinità : maschio e femmina, luce ed ombra, caldo e freddo.

10. Il rito di iniziazione al Waya, al quale ho preso parte presso il villaggio Xavante di San Gradouro, si è svolto il 31 luglio e il 1° agosto 1987. Non avendo potuto partecipare a quelle fasi del rito vietate alle donne, alcune informazioni mi sono state riferite da collaboratori.

questa, simbolo di morte. Ritengo che la voce dell'*upawá* voglia indicare che durante la cerimonia avviene un grande cambiamento : muoiono simbolicamente i giovani iniziandi e, al loro posto, nascono dei giovani che conoscono i segreti del *wayá*.

Tra gli strumenti indigeni più diffusi in America Latina e che in molti casi hanno quindi perso il loro significato rituale, troviamo flauti di diverso genere, tamburi (si pensi alla *tynia* delle Ande e al *teponaztli* o al *huehuetl* del Messico), raschiatoi di legno o di zucca e le maracas che vengono a ritmare, affiancati solitamente da strumenti di origine europea, ad esempio musiche e danze andine e pure brani venezuelani e colombiani tipici delle regioni costiere e delle pianure.

L'origine di altri strumenti musicali, come l'arco musicale e la marimba, non è ben chiara : il primo viene usato sia da indios (ad esempio i Miskitos dell'Honduras e gli Shuar dell'Ecuador), sia per accompagnare alcune danze e canti afro-brasiliani. La marimba è oramai parte fondamentale di molti complessi mesoamericani ; spesso viene suonata da popolazioni indigene come i Maya odierni sia per eseguire il rpertorio folklorico tipico della regione (repertorio cioè con forti caratteristiche europee), sia per accompagnare cerimonie e danze rituali tipiche del gruppo[11] ; ancora la marimba è uno strumento fondamentale durante l'esecuzione di brani afroecuadoriani della provincia costiera di Esmeralda[12] e, in questo caso, la sua origine è con evidenza africana.

A mio avviso non è da escludere che i due strumenti, l'arco musicale e la marimba, possano essere nati contemporaneamente nei due continenti, quello americano e quello africano. Ad esempio, a testimoniare l'origine amerindia della marimba, vi è un vaso maya del periodo quichés-tolteca dove è raffigurato un suonatore di marimba che trasporta lo strumento sulle spalle, proprio come ancora oggi viene trasportato dagli indios Maya-Quichés. Inoltre gli stessi indios non chiamano tale strumento marimba, bensì ojom, nome usato anche all'epoca della conquista[13].

Gli strumenti di origine africana

Questi sono presenti ovviamente dove maggiore è la concentrazione della popolazione negra o mulatta e cioè nelle Antille, nelle Guyane, sulle coste del Panama, Ecuador, Colombia, Venezuela, Brasile fino all'area del Rio della Plata.

11. G. Béhague, " Guatemala ", *Dizionario Enciclopedico della Musica e dei Musicisti*, II, Torino, 1983-1984, 450-451 (3 vols).

12. G. Béhague, " Ecuador ", *Dizionario Enciclopedico della Musica e dei Musicisti*, II, *op. cit.*, 101-102.

13. M. López Mayorical, " Origen Quiche de la marimaba ", *La antropologia americanista en la actualidad. Homenaje a Raphael Girard*, II, México, 1980, 107-110 (2 vols).

Molti di essi sono stati ricostruiti nel Nuovo Mondo con tutte le loro caratteristiche africane ; altri, invece, hanno subito delle trasformazioni adattandosi cioè al nuovo ambiente e a nuovi bisogni e necessità ; altri ancora sono di origine recente, sono cioè nati nel Nuovo Mondo sempre da popolazioni di origine africana, sotto l'impulso di necessità particolari. È il caso degli *steel drums* cioè dei bidoni metallici che caratterizzano le musiche di alcune isole delle Antille e che formano delle intere orchestre. Essi sono di origine assai recente e più precisamente del nostro secolo ; gli *steel drums* sostituirono infatti i *tamboo bamboo*, percussioni improvvisate di *bamboo* nate dal divieto imposto dai bianchi di usare strumenti appartenenti alla tradizione africana e che nel 1920 vennero nuovamente vietati dalla polizia in seguito a disordini scoppiati tra gruppi di percussionisti. Le percussioni metalliche risultarono comunque assai più sonore di quelle di bamboo e per questo le sostituirono[14].

Nonostante i vari divieti, infatti, i negri seppero sempre ricreare nuove situazioni musicali e nuovi strumenti. Molto spesso furono adottati tamburi bianchi dai quali ricavavano i loro particolari effetti percussivi[15].

Là dove i culti di origine africana sono ancora praticati, anche se spesso in sincretismo con elementi della religione cattolica, gli strumenti, pure essi di chiara origine africana, sono considerati sacri. Nel *candomblé* praticato a Salvador Bahia, a Recife e nel Maranhão, e nella *santeria* cubana, i tamburi vengono nutriti con sangue di animali e a volte pure con altre offerte, per trattenere la loro forza derivante dalle divinità.

Dove invece la religione cattolica ha preso il sopravvento, sono rimaste tracce rituali di origine africana evidenti nei ritmi e nell'uso di strumenti dalla stessa origine ; è il caso dei tamburi afrovenezuelani *mina* e *curbata* in uso in alcune regioni costiere del Venezuela, tamburi che accompagnano la processione e la festa che si svolge fuori dalla chiesa in onore di San Giovanni[16].

Tra gli strumenti afro-latinoamericani troviamo numerosi membranofoni, più precisamente tamburi generalmente ad una pelle percossi o con le mani o con bachette, e un gran numero di idiofoni ; tra questi i più comuni sono la sansa (zucca svuotata o scatola di legno su cui vengono applicate delle linguette metalliche sollecitate poi con le dita), alcuni raschiatoi come il *guayo* (crepitacolo di metallo in uso a Santo Domingo), sonagli soprattutto di metallo in uso durante i culti afro-latinoamericani, idiofoni a scuotimento non di metallo come il *piano-de-cuia* (sorta di maracas del Brasile avvolta in una rete di semi). Altri strumenti di origine africana sono l'arco musicale come il

14. S. Straatman, " Musik der Karib ", in C. Raddatz (a cura di), *Afrika in Amerika*, Hamburg, 1992, 200-201.
15. I. Leymarie, *Du tango au reggae. Musiques noires d'Amérique Latine et de Caraïbes, op. cit.*, 16-17.
16. *Du tango au reggae. Musiques noires d'Amérique Latine et de Caraïbes, op. cit.*, 45-46, 164-165.

carángano e il *berimbau* rispettivamente in uso in Venezuela e nel Nord-Est del Brasile, e la marimba in uso sulla costa dell'Ecuador.

Gli strumenti musicali di origine europea

Gli strumenti usati nelle nostre orchestre odierne di musica classica, sono ovviamente presenti oggi pure nelle grosse orchestre latinoamericane. Ciò che spesso è difficile definire, è l'arrivo di tali strumenti dall'Europa : non tutti infatti sono giunti nello stesso tempo. È comunque presumibile che gli europei si siano portati dietro, fin dall'epoca della conquista, parte degli oggetti appartenenti alla loro cultura e, con questi, anche alcuni strumenti musicali. I primi a raggiungere le nuove terre sono stati sicuramente gli strumenti a corde pizzicate come la vihuela e l'arpa diatonica (l'arpa cioè priva di pedali, come era in uso all'epoca della conquista in Europa), alcuni strumenti a fiato e a percussione indispensabili per l'esecuzione di musica militare[17] e l'organo, introdotto nelle chiese dei grossi centri urbani fin dal 1500.

Molti strumenti sono arrivati più tardi semplicemente perchè la loro origine è più tarda rispetto all'epoca della conquista. È ad esempio il caso di alcuni aerofoni (strumenti ad aria) come il clarinetto, costruito negli ultimi anni del 1600, e la fisarmonica che è del XIX secolo.

I conquistatori portarono quegli strumenti che erano più in uso nel loro paese e, più precisamente, nella penisola iberica. Per questo motivo l'uso della vihuela è documentato fin dal XVI secolo mentre il clavicembalo, strumento all'epoca probabilmente non tanto diffuso in Spagna e Portogallo giunse assi più tardi nel Nuovo Mondo : a Caracas, la prima documentazione risale al 1634[18].

Nel XIX secolo, con la fine dell'epoca coloniale, diversi paesi si sforzarono per promuovere attività artistiche di livello europeo, costruendo teatri e conservatori. Fu in questo periodo che si diffuse il pianoforte nell'alta società latinoamericana.

La fisarmonica e l'organetto arrivarono in America sempre nel XIX secolo, probabilmente con un'ondata di emigranti europei, solitamente gente assai povera che affrontava una nuova vita in cerca di fortuna. Infatti con l'indipendenza dei singoli stati latinoamericani, tutti gli europei poterono emigrare europea in America del Sud e Centrale (prima erano solo gli spagnoli e i portoghesi) e portarono seco nuove mode, ritmi e strumenti musicali.

È da ricordare come degli strumenti musicali europei si appropriarono velocemente e con una certa facilità sia i popoli autoctoni americani, sia quelli di origine africana. Come si è già detto in precedenza, i missionari intrapresero infatti una vera e propria educazione musicale rivolta agli amerindi. Secondo

---

17. V. Mariz, *Historia da música no Brasil*, Rio de Janeiro, 1994, 44-45.
18. A. Calzavara, *Historia de la Música en Venezuela*, Caracas, 1987, 53.

il gesuita spagnolo José Cardiel, che iniziò a lavorare in Paraguay attorno al 1730, ci riferisce che ogni villaggio aveva circa quaranta musicisti dei quali una parte erano cantori e una parte si dedicava invece alla pratica del violino, dell'oboe, della tromba, dell'arpa e dell'organo. Secondo il gesuita, molti di questi musicisti sarebbero stati considerati di eccellente livello anche nelle più grosse cattedrali europee[19].

Per quanto riguarda la facilità con cui gli africani si impadronirono degli strumenti musicali europei, riporto un ulteriore esempio : a Cuba già dal 1500 gli schiavi suonavano strumenti europei come la vihuela e l'arpa ; alcuni di essi eseguivano musica religiosa, altri animavano i balli organizzati dai loro padroni. Nei secoli seguenti dei musicisti negri suonavano nelle grosse orchestre e teatri e nel 1831 pare che, sempre a Cuba, i musicisti di colore fossero tre volte più numerosi di quelli bianchi[20].

Infine è da ricordare come di alcuni strumenti musicali europei si siano appropriati le popolazioni indigene, attribuendo loro una preciso significato all'interno di determinati rituali. Uno di questi strumenti è, ad esempio, il violino che, impoveritosi notevolmente dal punto di vista costruttivo (è in parole povere molto rozzo tanto che può essere difficile confrontarlo con un violino), si è invece arricchito di altri significati rituali. È il caso del *sekeseke*, violino degli indios Warao che vivono sul delta dell'Orinoco ; il *sekeseke*, introdotto dai missionari presenti in questa regione nel XVII secolo, è uno strumento la cui sacralità è evidente sia per la sua presenza nella mitologia di questi indios, sia perchè è parte fondamentale di determinati rituali[21].

Altro strumento europeo di cui si sono appropriati gli indios, è il tamburo *tu/tu* usato dai Tikuna dell'alto rio Solimões ; questo membranofono a doppia pelle e con corda di risonanza, è probabilmente stato introdotto dai portoghesi mediante le loro esercitazioni militari che si svolgevano nel forte di Tabatinga nel XVII secolo[22] ; oggi il tamburo *tu/tu* viene usato durante il rito di iniziazione denominato della *moça nova* ; su una delle sue pelli vengono spesso disegnati col fuoco, animali ben precisi, come un uccello o un giaguaro, che simboleggiano un clan di appartenenza dell'iniziando. Il suono del tamburo accompagna con insistenza tutto il rituale ; vuole forse aiutare i partecipanti alla cerimonia a raggiungere quello stato di *trance* necessario per raggiungere il contatto con gli spiriti superiori[23].

19. G. Béhague, *La musica en America Latina*, op. cit., 23-24.

20. I. Leymarie, *Du tango au reggae. Musiques noires d'Amérique Latine et de Caraïbes*, op. cit., 26.

21. I. Aretz, *Música de los aborigenes de Venezuela*, Caracas, 1991, 282.

22. C. Nimuendajú, *The Tukuna*, Berkeley, Los Angeles, 1952, 42-46.

23. A. Minelli, C. Minelli, *Amazzonia. Là dove gli alberi sostengono il cielo*, Bologna, 1992, 254.

Ancora ricordo come alcuni strumenti musicali di origine europea si siano trasformati nel Nuovo Mondo, adattandosi a quelle realtà musicali locali ricche di sincretismi culturali ; è il caso del *charango*, strumento a corde pizzicate oggi comune in Perù, Bolivia e Cile e che è il risultato di un adattamento popolare della vihuela.

SCALE, MELODIE E CANTI

A prescindere da quelle musiche tipicamente indigene rintracciabili con evidenza presso quei gruppi amerindi rimasti e che vivono soprattutto nella foresta tropicale, la musica indigena ha influenzato, in maniera più o meno rilevante, il folklore di alcune regioni latinoamericane.

Vi sono paesi dove gli indios vivono da tantissimo tempo a contatto con la civiltà occidentale, acquisendo canti e danze dalle origini europee che si sono affiancati a quelli autoctoni. Il Messico, il Guatemala e, in maniera ancora più evidente, la regione andina sono esempi dove si riscontrano le due culture, quella indigena e quella europea, coesistenti e affiancate ; per essere più chiari, gli indios suonano, cantano e ballano sia il loro antico repertorio, sia quello di origine europea.

Sulle Ande sono infatti presenti ancora oggi canti e danze di origine preispanica strutturati su scale pentatoniche come lo *yaraví*, canto d'amore dall'andamento malinconico, o il *huayno*, danza vivace dal ritmo sincopato, o ancora la *baguala* argentina, melodia lenta strutturata invece su una scala tritonica. Accanto a questo repertorio, gli stessi musici suonano e cantano canti e danze di origine spagnola appartenenti ad esempio al gruppo della *zamacueca*[24] e, in Argentina, canti modali la cui origine risale all'epoca della conquista[25].

Ancora, gli indios Makuxí di Roraima (Brasile), cantano, accanto al loro antico repertorio rituale strutturato su semplici melodie generalmente dall' andamento discendente, canti per lo più appartenenti al reportorio della religione cattolica insegnati e appresi con facilità, dai missionari che lavorano in questa regione. Alcuni di questi canti, per l'esattezza quelli appartenenti al culto dell'*alleluia*, risalgono al XIX secolo e sono la testimonianza di uno di quei tanti sincretismi religiosi presenti in America Latina[26].

Alcuni indios, ad esempio gruppi che vivono in Messico, posseggono melodie del loro repertorio amerindio che richiamano, per la loro struttura modale

24. G. Béhague, " Perù ", *Dizionario Enciclopedico della Musica e dei Musicisti*, III, *op. cit.*, 628-629.

25. G. Béhague, " Argentina ", *Dizionario Enciclopedico della Musica e dei Musicisti*, I, *op. cit.*, 11-114.

26. Le notizie qui riportate sugli indios Makuxí, sono frutto sia di una mia esperienza sul campo avvenuta in agosto del 1996, sia di informazioni avute dai missionari che lavorano tra questo gruppo di indios.

e ritmica, il canto gregoriano o anche antiche melodie modali europee ; un esempio sono i canti sciamanici degli indios Mazatec di Oaxaca[27].

In questo caso gli indios, come già abbiamo visto per alcuni strumenti musicali, si sono appropriati di elementi europei : i canti appartenenti ad esempio al repertorio gregoriano, insegnati a loro dai missionari all'epoca della conquista o negli anni subito seguenti, hanno oggi assunto un valore rituale indigeno.

Padre Josè de Acosta nella *Historia Natural y Moral de las Indias* pubblicata per la prima volta a Siviglia nel 1590, ci porta una testimonianza di come arrivarono tali canti in America Latina : i missionari evangelizzarono sfruttando la propensione e il gusto dei popoli amerindi verso la musica ed il canto ; a volte vennero ripresi i canti autoctoni, altre volte si tradussero, invece, in lingua indigena *octavas, canciones, romances* e *redondillas*, canti che gli indios impararono con incredibile rapidità gustandone appieno la bellezza[28].

Nelle regioni dalla forte influenza africana sono riscontrabili parecchie caratteristiche che affondano le proprie radici nel continente nero come le melodie dall'andamento discendente, il canto responsoriale, le scale pentatoniche o diatoniche con il settimo grado bemolizzato o addirittura assente[29].

Ma il vero sopravvento sulle culture appena citate, cioè quella africana e quella indigena, è pervenuto dalla cultura europea : per essere più chiari, la tonalità oggi predomina su tutti gli altri linguaggi musicali. Così molte danze la cui origine africana è evidente nella coreografia o nel ritmo, hanno oggi assunto melodie tonali.

Scale e melodie amerindie o africane, così come tutta la musica di queste culture, sono rimaste intatte, soprattutto là dove esistono le religioni originarie. Dove invece la musica ha perso il suo significato rituale, maggiore è stata l'influenza della cultura europea.

RIFLESSIONI SUI RITMI E ALCUNE DANZE LATINOAMERICANE

Sicuramente noi europei abbiamo una concezione e un modo di sentire il ritmo diverso dai latinoamericani. Mi sono sempre chiesta a tale proposito, perchè all'interno delle scuole dell'obbligo, ma anche delle scuole di musica e dei conservatori, risulta difficile insegnare agli alunni ad eseguire figure ritmiche più o meno complesse. Al contrario i latinoamericani, fin da bambini, sembra abbiano un senso innato nell'eseguire non solo correttamente ma pure con una certa vivacità, brani poliritmici e ciò senza saper leggere una nota o un valore sullo spartito.

27. G. Béhague, " Messico ", *Dizionario Enciclopedico della Musica e dei Musicisti*, III, *op. cit.*, 126-129.

28. J. de Acosta, *Historia Natural y Moral de las Indias*, Madrid, 1954, 207.

29. A. Vasconcelos, *Raízes da música popular brasileira*, Rio de Janeiro, 1991, 22.

Il perchè di queste differenze sta, a mio avviso, ancora una volta nella storia dei due mondi e del bagaglio culturale che ciascuno si è portato dietro e ha poi continuamente trasformato secondo necessità diverse.

In primo luogo in Europa, come si è già detto precedentemente, è nata e si è sviluppata la scrittura musicale. Saper mettere sulla carta la musica significa anche poterla elaborare sviluppando diverse caratteristiche musicali quali la melodia, il ritmo, l'armonia, la ricerca timbrica. Nella musica colta occidentale, il ritmo si è quindi spesso trovato a far parte di un equilibrio tra elementi diversi, voluto dal compositore. Il ritmo può quindi essere messo in evidenza o, invece, rimanere nascosto dietro ad altre caratteristiche del brano (la cantabilità, la ricerca timbrica...).

Ancora, la storia della musica colta occidentale è legata sia alle pulsazioni ritmiche della danza, sia alla parola. In quest'ultimo caso il linguaggio musicale e verbale sono strettamente connessi e i loro ritmi sembrano fondersi. Così, ad esempio, nel gregoriano il ritmo del canto dipende da quello delle parole e delle frasi testuali[30]. Secondo alcuni studiosi come Van der Werf anche il ritmo della maggior parte delle canzoni trobadoriche e trovieriche dipendeva dal fluire e dal significato del testo[31]. Con l'avvento della polifonia fu necessario definire un sistema di valori che regolasse la durata delle note e che concettualmente derivarono dalla metrica e quindi, ancora una volta, dalle discipline del linguaggio[32].

Tutto ciò sta a significare quanto importante fosse in Europa, fin dai tempi antichi, il legame tra parola e musica, legame che poi si è sempre mantenuto pur trasformato dal mutare e dall'evolversi dei tempi e quindi anche delle arti e che nella storia della musica moderna, si trova ben visibile nell'opera e nella musica sacra, così come nella musica leggera. Sicuramente anche la danza di derivazione popolare, con i suoi ritmi più incisivi ed energici, ha influenzato la musica colta e più precisamente quella strumentale entrando a far parte, in maniera stilizzata, di composizioni raffinate e geniali dove, come si diceva in precedenza, sono entrati in equilibrio elementi diversi.

Colto e popolare si sono, nella storia della musica occidentale, in continuazione influenzati vicendevolmente. Anche nel campo della musica popolare troviamo infatti i due filoni, quello legato al linguaggio parlato (ad esempio il canto narrativo) e quello invece legato alla danza.

Completamente diversa è la concezione del ritmo sia per le popolazioni amerindie, sia per quelle provenienti dal continente africano : la mancanza di scrittura non ha sicuramente permesso di sviluppare certe caratteristiche come il gioco delle parti o l'armonia, tipiche della nostra musica colta, ma ha forse dato più incisività ad altri elementi, primo tra tutti il ritmo anche perchè soli-

---

30. G. Cattin, *Il Medioevo*, I, Torino, 1981, 79-84.
31. *Idem*, 153.
32. A. Gallo, *Il Medioevo*, II, Torino, 1981, 3-11.

tamente associato alla danza ; inoltre, ancora una volta, entra in gioco la sacralità della musica.

Presso la cultura popolare afro-latinoamericana, il ritmo assume infatti un ruolo assai rilevante e di primo piano : associato sempre alla danza, ne è l'impulso vitale. Le percussioni hanno infatti una parte realmente primaria in molte danze dove il supporto armonico può non sussistere o è assai povero così come la melodia può essere espressione di un solo strumento, come la chitarra o il cavaquinho[33], e della voce. È il caso del samba di carnevale brasiliano la cui reale forza nasce dall'iterazione dei vari ritmi ottenuti da diversi strumenti a percussione e, sopra di essi, una voce canta una melodia. Il ritmo è, in questo caso, travolgente e chi balla, viene a mio avviso a cadere in uno stato di semiincoscienza.

Nei rituali di origine africana questa forza del ritmo viene ad essere ancora più evidente : il suono dei tamburi richiama infatti le divinità ; il ritmo è quindi sacro e chi partecipa al rito, " cavalcato " da una divinità, balla cadendo in *trance*. Di ciò mi sono resa conto non solo consultando fonti bibliografiche ma anche partecipando, come spettatrice, a rituali *umbanda*[34] e a un rito *macumba*[35].

Presso le culture afro-latinoamericane, inoltre, i tamburi non hanno solo uno scopo ritmico : esprimono frasi musicali complete cariche anche di linee melodiche. I batteristi ricavano cioè dal proprio strumento non solo ritmi più o meno complessi ma, percuotendo i tamburi in punti diversi, anche suoni dalle altezze differenti che quindi vengono a formare una melodia[36]. I tamburi sono in questo caso strumenti a percussione che noi europei chiameremo a suono determinato : essi sono quindi strumenti ritmici, melodici e sacri[37].

Anche presso gli indios si trova sia l'unione tra musica e danza, sia la sacralità della musica che investe naturalmente anche gli strumenti ed il ritmo : musica e danza si caricano cioè di significati e simboli rituali estranei al nostro mondo occidentale moderno. Così ad esempio strumenti idiofoni come i sonagli o i bastoni sonaglio, possono servire allo sciamano sia a scacciare spiriti nocivi all'uomo (ad esempio durante i riti di guarigione), sia a mettersi in contatto con spiriti che vivono nel mondo di sopra o di sotto ; è il caso del bastone sonaglio usato dal maestro di cerimonia durante il rito della moça nova degli indios Tikuna, bastone che serve a richiamare, percuotendo il terreno, gli spiriti che vivono nel mondo di sotto e di sopra. La monotonia di certi ritmi, può inol-

---

33. Il cavaquinho è una piccola chiatarra con quattro corde metalliche di origine portoghese e in uso in Brasile. A Cabo Verde e, seppur con un altro nome, nelle Hawai e in Indonesia.

34. Belém, 1994 ;, Poço de Caldas, 1995.

35. Rio, 1995.

36. O. Calderón, " Las bases ritmicas afro : algunos errores de apreciación ", in I. Aretz (a cura di), *Anuario FUNDEF*, anno I, Caracas, 1990, 63-66.

37. La stessa sacralità della musica così come la sua unione con la danza, la si trova anche presso le antiche civiltà precolombiane.

tre aiutare a cadere in trance permettendo quindi il contatto con gli spiriti superiori.

Non esistendo la scrittura musicale presso le culture amerindie e di origine africana, la musica e quindi anche il ritmo, vengono tramandati oralmente. Il loro apprendimento non coinvolge semplicemente la mente e la memoria ma coinvolge, a mio avviso, anche lo stretto legame che hanno con il movimento del corpo (musica e danza sono generalmente associate) : le pulsazioni del ritmo sembrano nascere dall'interiorità di ciascun esecutore e sono rivolte proprio ai danzatori e al loro sfrenato movimento.

Molte sono le danze, soprattutto di origine africana, che hanno oggi perso il loro significato rituale o che comunque si sono caricate di elementi bianchi. Isabelle Leymarie cita il *candombe* uruguayano, divenuto spettacolo di carnevale, il *tango* oggi di dominio della borghesia argentina, e ancora il *samba* brasiliano e la *rumba* cubana nella cui coreografia sono presenti caratteristiche di antichi riti di fertilità bantù[38].

A questo riguardo Artur Ramos ricorda come il *quizomba*, danza nunziale angolana giudicata dagli europei indecorosa a causa della sua coreografia, abbia fortemente influenzato i *batuques* e i *sambas* brasiliani[39]. E infatti oggi il *samba* ha mantenuto, pur essendo un ballo profano, dei forti significati erotici e gli stessi ritmi travolgenti che sembrano far cadere in uno stato simile al *trance* chi lo balla, dimostrano chiaramente, a mio avviso, le sue radici rituali.

Le danze latinoamericane sono un universo sempre pronto ad evolversi, ad acquisire e a trasformare elementi nuovi provenienti da altri paesi sia europei che americani ; così forse si possono sintetizzare i cinque secoli trascorsi dopo la conquista, secoli che hanno visto decadere, dimenticare, trasformare e creare in continuazione decine di danze. Oggi alcune di esse conservano una chiara matrice europea, evidente soprattutto nella coreografia, nelle scale usate e pure nei ritmi.

Nei paesi ispanoamericani, infatti, moltissimi sono i balli dai ritmi sovrapposti di 6/8 e 3/4, così come coreografie che fanno uso di fazzoletti o del battere ritmico dei tacchi. A volte hanno acquisito elementi locali o di paesi vicini ; è il caso di parecchie danze dell'America Centrale o Meridionale come il *jarabe* messicano, il *son guatemalteco*, il *punto guanacasteco* e la *callejera* di Costarica, la *cueca* cilena e argentina, la *zamba* e la *chacarrera* argentina.

Spesso è difficile distinguere la provenienza dei ritmi, la loro evoluzione e trasformazione. Un esempio può essere, a mio avviso, il *joropo* venezuelano, danza che presenta una poliritmia derivante da una differente accentuazione delle parti. Secondo Isabelle Leymarie la sua origine è spagnola anche se a

---

38. I. Leymarie, *Du tango au reggae. Musiques noires d'Amérique Latine et de Caraïbes, op. cit.*, 18.

39. A. Ramos, *O negro brasileiro, op. cit.*, 150-151.

Barlovento e nella vallata del Tuy il ritmo sincopato attesta un'influenza negra[40].

Nei paesi di maggior presenza africana, molte danze di origine iberica hanno acquisito ritmi sincopati e un'esecuzione percussiva degli strumenti europei, elementi che sono evidenti, ad esempio, in molti balli cubani.

Molti ritmi e danze provenienti dall'Africa, come abbiamo già visto, vengono oggi eseguiti in feste cattoliche ; la stessa cosa è accaduta per alcuni balli rituali amerindi di cui abbiamo già ricordato l'*alleluia* danzato nel periodo natalizio e in uso presso i Makuxí del Brasile ; Gerard Béhague cita altri esempi di cristianizzazione di danze rituali autoctone nella regione andina[41] e Oneyda Alvarenga, parlando dell'origine del *Cateretê* in uso soprattutto negli stati di S. Paulo, Rio, Minas Gerais, Mato Grosso e Goiás del Brasile, ci dice che " secondo l'opinione corrente, è danza amerindia, utilizzata nel primo secolo della colonizzazione da Padre Anchieta nelle feste cattoliche, in cui la si danzava e cantava con testi cristiani, scritti in dialetto tupi "[42].

Le danze europee non sono giunte in America Latina solo all'epoca della conquista o poco dopo; le mode europee hanno sempre raggiunto il Mondo Nuovo e, con queste, anche i balli in voga. Così è successo con la polka, la mazurka e il valzer viennese. Il valzer, in modo particolare, è penetrato quasi ovunque, acquisendo spesso nuovi elementi locali. Ad esempio in Venezuela, il valzer conquistò i compositori di musica colta della fine dell' '800 le cui composizioni sono caratterizzate spesso da ritmi in 6/8 e 3/4 sovrapposti. In Colombia il valzer si è trasformato in *pasillo*, una danza in forma di rondò dall'andamento vivace che, a cavallo dell'ultimo secolo, conquistò, così come il valzer venezuelano, anche i compositori di musica colta. In Ecuador il *pasillo* ha invece un andamento più lento e può essere sia cantato che esclusivamente strumentale. Anche la *guaranía* del Paraguay è in tempo di valzer lento ma è la *polca paraguaya* il genere musicale che ha avuto più fortuna in questo paese : derivato appunto dalla polka europea, conserva la melodia in ritmo binario vivace, mentre ha l'accompagnamento generalmente in ritmo ternario. In Brasile, secondo José Ramos Tinhorão, il *maxixe* nato alla fine del 1800 come danza e poi sopravvissuto solo come genere di canzone, nasce dall'adattamento popolare delle danze allora in voga come, appunto, la *polka*, lo *scottissch* e la *mazurka*[43].

La vitalità musicale, l'amore, la passione per la musica e la danza fa si che in America Latina gli stili e i ritmi si spostino velocemente da un paese

---

40. I. Leymarie, *Du tango au reggae. Musiques noires d'Amérique Latine et de Caraïbes*, op. cit., 169-170.

41. G. Béhague, " Ecuador ", *Dizionario Enciclopedico della Musica e dei Musicisti*, II, op. cit., 101-102.

42. O. Alvarenga, *Musica popolare brasiliana*, Milano, 1953, 163-166.

43. J. Ramos Tinhorão, *Pequena história da música popular*, S. Paulo, 1991, 58-96.

all'altro creando spesso occasioni per la creazione di nuovi generi e danze :
uno degli esempi più significativi è, a mio avviso, quello del *tango* argentino
il cui ritmo è stato fortemente influenzato dall'*habanera* cubana portata in
Argentina da marinai dai Caraibi nel 1850[44].

Infine voglio ricordare come l'urbanizzazione di certe danze tradizionali
soprattutto di origine africana, ha portato alla perdita di elementi coreografici
e percussivi di chiara derivazione negra, per acquisire melodie e armonie più
affini all'orecchio comune delle città. Ricordo ad esempio la *cumbia* colom-
biana, la *rumba* cubana, il *tamborito* del Panama, il *merengue* dominicano,
danze e ritmi che, diventando una moda sono inoltre spesso usciti dal proprio
paese per essere eseguite e ballate anche negli stati vicini.

Alle danze e ai ritmi popolari si sono ispirati moltissimi compositori latinoamericani di musica colta, compositori appartenenti a quella corrente nazio-
nalista che in questo continente è stata molto viva e sentita fino circa al 1950[45].
Compositori come Heitor Villa-Lobos, Alberto Ginasteira, Astor Piazzolla,
Camargo Guarnieri, Radamés Gnattali, Guerra Peixe, per fare il nome solo di
pochi, si sono spesso ispirati a un folklore ricco e vivo, trasformandolo in
musica colta spesso impregnata di idee originali e all'avanguardia.

Con questa riflessione sulla musica latinoamericana, che prende spunto
soprattutto dalla musica popolare e tradizionale, ho voluto mettere in risalto
come dall'incontro di tre culture diverse, quella indigena, quella europea e
quella africana, si sia formato un mondo musicale estremamente vitale, ricco
di sincretismi e, ancora oggi, in continua evoluzione. Tale incontro di culture è
iniziato, in maniera più o meno violenta, dopo la conquista del Nuovo Mondo
ad opera degli europei ; ma i sincretismi e l'evoluzione musicale di questo continente non si fermano nel 1500, bensì continuano nei secoli seguenti, alimen-
tate dalla stessa storia dell'America Latina, dal commercio di schiavi africani,
dall'imposizione della cultura europea che nella maggior parte dei casi ha ten-
tato di schiacciare le culture più deboli, dalle migrazioni avvenute nei secoli
seguenti sia all'interno dell'America stessa, sia dall'Europa come pure da altri
continenti.

Dall'incontro di queste tre culture sono nati degli ibridismi musicali che,
come si è già messo in evidenza precedentemente, sono diversi da regione a
regione, e dipendono dalle culture che vi predominano. Così ad esempio, il
merengue dominicano presenta ritmi enfaticamente sincopati e irregolari di
chiara influenza africana, scale europee e strumenti musicali derivati o appartenenti a una delle due culture come la fisarmonica (cultura europea) e un tipo

---

44. I. Leymarie, *Du tango au reggae. Musiques noires d'Amérique Latine et de Caraïbes*, *op.
cit.*, 222-223. Il *tango*, nasce nei sobborghi di Buenos Aires alla fine del 1800 da elementi di origine africana, dal *tango andaluso*, dalla *milonga* (genere cantato e succesivamente danzato che
faceva parte del repertorio del *payador* argentino cioè del cantante specializzato in tenzoni
canore), dall'*habanera* cubana portata in Argentina da marinai dai Caraibi nel 1850.

45. G. Béhague, *La musica en America Latina, op. cit.*, 183.

di sansa denominato marimbula (cultura africana). Ancora, in gran parte delle regioni andine i cordofoni, provenienti dal Vecchio Mondo, affiancano strumenti autoctoni come i vari tipi di flauti o la *tinya*, piccolo tamburo di origine precolombiana.

Tale mondo musicale è ancora oggi estremamente vitale, dando vita a generi e correnti nuovi come la *salsa* dei Caraibi, nata a New York dagli emigranti cubani e portoricani, o la *bossa nova*, nata in Brasile dall'apertura verso il mondo del jazz, o ancora la *lambada* brasiliana, sorta dal *carimbò* (genere di *batuque* tipico del Pará brasilino) e da ritmi afro-caribici[46]. Fondere elementi nuovi significa prima di tutto essere recettivi verso il mondo della musica e, insieme, essere sensibili verso altri paesi e culture diverse.

Ancora, la vitalità musicale latinoamericana crea generi che molto spesso si spingono molto oltre la semplice musica popolare, verso un confine indefinito con il colto (si pensi ai valzer venezuelani della fine del 1800 o a certi tanghi argentini di oggi). Infine vitalità che ancora oggi, seppur in maniera molto più attenuata rispetto alla prima metà del secolo, crea continui contatti a doppio senso tra popolare e colto e dove inevitabilmente il genere povero, proprio perchè così radicato nella cultura, non può non influenzare, anche se in maniera più o meno rilevante, la musica erudita.

L'evoluzione e la storia della musica latinoamericana è impregnata di un modo di intendere e di fare musica che ancora oggi è molto sentito per cui si impara a suonare e a ballare fin da bambini, il più delle volte senza conoscere il pentagramma e la scrittura musicale. La musica mantiene infatti, in America Latina, una sua ragione d'essere sia per le sue tradizioni religiose ancora vive da cui in parte proviene, sia perchè è considerata, assieme alla danza, uno dei principali divertimenti[47], sia perchè ancora non è stata schiacciata dal mondo consumistico ed è quindi una tradizione molto viva.

Questo può forse spiegare la musicalità di quei bambini che incontrai in pieno centro a Caracas e che, con un vecchio violino, due maracas, un cuatro e una piccola arpa diatonica, suonavano intonatissimi la poliritmica danza locale denominata *joropo*. Nessuno di loro sicuramente sapeva leggere la musica e, come loro, tanti altri musicisti piccoli e grandi che suonano per le strade. Nelle nostre scuole a fatica i ragazzi imparano i primi elementi del flauto dolce e nei nostri conservatori spesso uno dei grandi problemi dell' apprendimento, è costituito dal ritmo.

L'incrocio tra tre culture e la loro evoluzione storica ha portato i musicisti latinoamericani quindi a una capacità nell'improvvisare e a un senso del ritmo che noi non possediamo. Gli stessi musicisti di formazione classica posseggono tali capacità, oltre ad avere ottimi studi per quanto riguarda la musica

---

46. J. Ramos Tinhorão, *Pequena história da música popular*, *op. cit.*, 278-291.

47. Anche il legame tra musica e danza è molto radicato in America Latina : lo si trova sia presso le culture precolombiane, sia presso gli attuali amerindi e afro-latino-americani.

colta e a saper dominare il proprio strumento spesso in maniera eccezionale. Molti di loro ci potrebbero sicuramente insegnare molto nel campo della musica.

Forse l'associazione tra ritmo e danza facilita l'apprendimento dei ritmi più complessi ? Forse il ritmo non è quindi qualcosa che sta scritto matematicamente sulla carta ma è qualcosa di molto più vitale ? Forse il fatto di suonare richiamando in continuazione la memoria di linee melodiche e ritmi (cioè senza spartito) aiuta a imparare l'arte dell'improvvisazione ?

Queste mie domande sicuramente nascono dalle mie esperienze di insegnante e di musicista così come sono il frutto di riflessioni sulla musica latinoamericana odierna, su quel mondo tanto vitale e spontaneo che ho in piccola parte conosciuto. Di quel mondo mi appassionano I significati e i simboli impregnati di religiosità della musica amerindia, i ritmi di origine africana così vivi e spesso carichi di forze sovrannaturali, la musicalità di un popolo considerato culturalmente povero e appartenente al terzo mondo, la capacità impressionante dei gruppi di musicisti colti che ho conosciuto, di improvvisare e di dominare il proprio strumento. Forse noi del Vecchio Mondo, con il nostro eurocentrismo e un modo di pensare a volte chiuso e rigido, potremmo aprirci a ciò che è diverso e che va oltre la pura musica di consumo, per arricchire il nostro mondo musicale : così hanno fatto i latinoamericani nel corso degli ultimi secoli, traendone solo vantaggi.

CONTRIBUTORS

Jean C. BAUDET
Bruxelles (Belgique)

Hannelore BERNHARDT
Berlin (Germany)

Sébastien BRUNET
Université de Liège
Liège (Belgique)

Martin COUNIHAN
University of Southampton
Southampton (United Kingdom)

Roberto DOATI
Università di Padova
Albignasego (Italy)

Roman DUDA
Universytetu Wrocławskiego
Wrocław (Poland)

Don FAUST
Northern Michigan University
Marquette, MI (USA)

Jean GADISSEUR
Université de Liège
Liège (Belgique)

Stephen GAUKROGER
University of Sydney
Sydney (Australia)

Yves GINGRAS
Université du Québec
Montréal (Canada)

Oleg I. GUBIN
The University of Utah
Salt Lake City, UT (USA)

Antonello LA VERGATA
Università della Calabria
Arcavacata (Italy)

Kurt LOCHER
University of Bern
Bern (Switzerland)

Ernest MATHIJS
Vrije Universiteit
Brussels (Belgium)

Carla MINELLI
Lisboa (Portugal)

Bert MOSSELMANS
Universiteit Antwerpen
Antwerpen (Belgium)

Peeter MÜÜRSEPP
Tallinn Pedagogical University
Keila (Estonia)

Mario H. OTERO
Universidad de la República (FHCE)
Montevideo (Uruguay)

Amalia PERFETTI
Università degli Studi di Firenze
Firenze (Italy)

Andrés RIVADULLA
Universidad Complutense de Madrid
Madrid (Spain)

Nicola SCALDAFERRI
Università di Bologna
Bologna (Italy)

Wolfgang von STROMER
Altdorf (Germany)

Alvise VIDOLIN
Conservatorio " B. Marcello "
Venezia (Italy)

Witold J. WILCZYŃSKI
WSP Institute of Geography
Kielce (Poland)

Martin ZERNER
Université de Nice
Paris (France)